SIX ACES

AUSTRALIAN FIGHTER PILOTS 1939-45

LEX McAULAY

First published 1991 by Banner Books, 308 Victoria Street, Brunswick, Melbourne, Vic., 3056.

Copyright © 1991 Lex McAulay

Cover design by David Constable, Carlton, Vic.
Text design by Kenneth Yendell, North Fitzroy, Vic.
Typeset in 13/14pt Bembo by Solo Typesetting, South Australia.
Printed in Australia by Impact Printing, Brunswick, Vic.

This book is copyright. Apart from any fair dealing for the purposes of private study, research, criticism or review as permitted under the Copyright Act, no part may be reproduced by any process without prior written permission from the publisher.

Cataloguing-in-publication information

McAulay, Lex, 1939- .
    Six aces: Australian fighter pilots, 1939-45.

    Bibliography.
    Includes index.
    ISBN 1 875593 00 4.

    1. Australia, Royal Australian Air Force—History—World War, 1939-45. 2. Fighter pilots—Australia—Biography. 3. World War, 1939-45—Aerial operations, Australian. I. Title.

940.544994092

Banner Books is Australia's specialist publisher of military history, committed to publishing high quality books which add a significant dimension to the history of Australians at war.

Also available:

The Decisive Factor—75 & 76 Squadrons—Port Moresby and Milne Bay 1942 by David Wilson.

If you would like to know more about Banner Books write to us at 308 Victoria Street, Brunswick Vic., 3056, Australia.

Contents

Acknowledgements		vii
Introduction		ix
That brand of courage	*Leslie Clisby*	1
Bader's bus company	*Tony Gaze*	59
The thin red line	*Jack Storey*	129
Never appear on my CO's parades!	*Mervyn Shipard*	157
Take that off my tax!	*Butch Gordon*	205
We never met his equal	*Charles Scherf*	239
Pulled up to clear the mast	*Max Cotton*	285
Footnotes		340
Bibliography		341

Acknowledgements and sources

As may be expected in such a work, many people gave their time, loaned personal items, and assisted with part of the research. As usual, the staff of the Australian War Memorial were most helpful, and I wish to thank Ian Affleck, Joyce Bradley, Bill Fogarty, George Imashev, Andrew Jack, Tony Rudnicki, Ian Smith and Pam Ray. At RAAF Historical Section, Jan Beck, Bob Piper and David Wilson were most supportive and knowledgeable. Wendy Gorton drew the maps.

Other people who assisted greatly were: Ing. Horst Amberg, Germany; Tom Anderson, Canada; General Jacques Andrieux CR DFC, France; C. Perry Arnold, Canada; Bernard Baeza, France; Victor Bingham, UK; Douglas Cotton, Australia; James P. Coyne, Canada; Brian Cull, UK; George Drew, UK; Mary (Clisby) Eckert, Australia; H. Frhr von Friesen, Germany; Tony Gaze DFC* RAF, Kev Ginanne and Russell Guest, all of Australia; Mrs E. Halahan, UK; James R.F. Johnson, Canada; Ron Jordan, John Howell and Dr Hugh Kennare AM, all of Australia; P.B. 'Laddie' Lucas CBE DSO DFC RAF, UK; Beryl (Scherf) McLachlan, L. Morrison, all of Australia; Douggie Oxby DSO DFC DFM* RAF, Canada; Hope Scherf, Mervyn Shipard DFC RAAF, all of Australia; Christopher F. Shores, UK; Mary Stewart, Canada; Jack Storey DFC MID RAAF, Australia; John Vasco, UK; Tom Warren, Australia; Phil Wigley DFC RAF, UK. My personal thanks to each and every one, especially to Phil Wigley, for his cheerful work at PRO Kew, and readiness to assist at any time.

Major sources for the chapters were:
Leslie Clisby: letters to family and biographical notes from Mary Eckert, Victorian Office of Australian Archives, 1 Squadron War Diary, via PRO Kew; crash reports: Australian Archives, Victoria. Accession No. P187/4, Item 161 'A7-45 Moth'; AWM 220.

Max Cotton: 263 Squadron war diary; flying log book, personal journal; AWM 220, Item 47 for information on RAF operations over Europe 1943, combat reports.

Tony Gaze: Flying log book, squadron war diaries, interviews and letters; AWM 220.

Reg 'Butch' Gordon: 31 Squadron Association, squadron war diary, combat reports, interviews and letters from Ron Jordan.

Charles Scherf: family information from Beryl McLachlan and Hope Scherf, squadron histories, from 418 Squadron RCAF Association; Information Services, Department of National Defence, Ontario, for the squadron war diary; flying log book; combat reports.

Mervyn Shipard: Flying log book, combat reports, squadron war diaries, interviews, letters.

Jack Storey: Flying log book, combat reports, interviews, letters.

Permission to use information in the following is gratefully acknowledged:
Macdonald & Company, London, for *Twice Vertical*, by Michael Shaw, 1971.

Permission to use the photos of Leslie Clisby and the crashed aircraft near Werribee was given by The Herald and Weekly Times Limited, and is gratefully acknowledged.

Introduction

Australian fighter pilots, as members of the RAF and RAAF, flew operations from almost the first to the last day of the Second World War. Many RAAF aircrew, not only fighter pilots, flew with non-Australian squadrons of the Commonwealth air forces as part of what may be the greatest gathering of many nationalities of volunteers for the dangerous duty of wartime flying the world has ever seen or will see. The careers of the six fighter aces described here cover the war theatres from France in 1940, through Burma, Malta and North Africa, the Darwin area, the English Channel, France again and through to Germany in the last days of the war, when jets were coming into service. Max Cotton, who flew anti-shipping strikes in the English Channel, is included as a representative of the many pilots who served to the best of their ability, and now are remembered by their squadron friends and families.

The aircraft flown by these men have all been the subjects of quite good and detailed books, so there is little point in repeating the technical descriptions here. Any reader wishing to know more about the fighters concerned can easily find such information elsewhere.

Six Aces is dedicated to the memory of all those RAAF volunteers who did not return, or who have passed on since 1945, and especially to the memory of my uncle, Tommy Berndt, killed in an aerial collision in 1943.

That brand of courage

Blitzkrieg Hurricane

Flg Offr Leslie R. Clisby DFC

SIX ACES

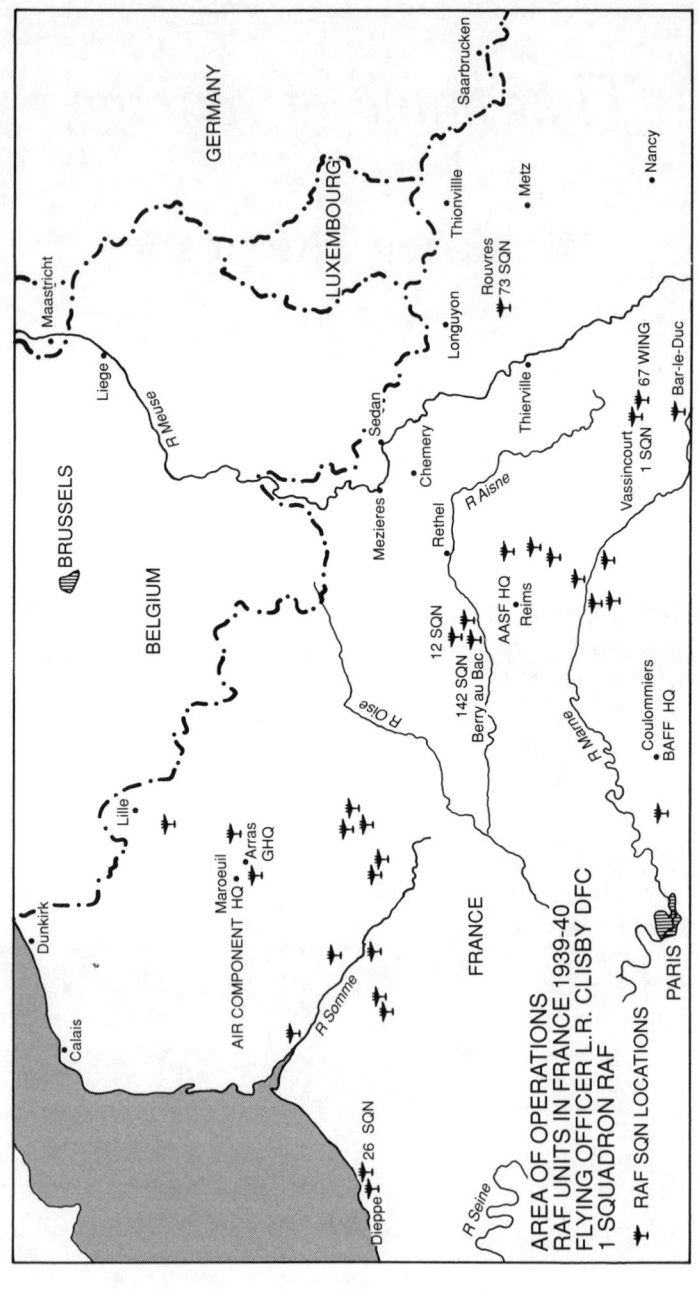

Leslie R. Clisby DFC

Leslie Redford Clisby was born on 29 June 1914 at McLaren Vale, South Australia.

Europe was sliding into war after the assassination of the Archduke Ferdinand at Sarajevo, and 25 years later Les Clisby would fly in the second great war, which grew from the wreckage of the first.

In 1919, the Clisby family moved to Walkerville, a suburb of Adelaide, where Les lived until he joined the RAAF at the age of 21. He was not a brilliant student, but did have determination to apply himself until the subject was mastered. One school report said that he was 'naturally slow, but his thoroughness and earnestness in everything in connection with his work accounts for its excellence.' Clisby's dedication resulted in his gaining second place in his final examinations and in the year's work.

Unusually for a mechanic and pilot, Les Clisby was left-handed. It was perhaps more unusual that he passed through school in those years as a left-hander, as conventional wisdom was for teachers to force the child to adapt to 'normal methods' of writing and using implements, despite the stress this inflicted on some unfortunate children. It was only some 20 or 25 years later that left-handed children were to be allowed to progress naturally. When young Clisby left school, his first job lasted only one day, when the employer discovered he was left-handed.

More typically of the young Aussie of the times, Clisby was enthusiastic about sport. He played cricket and football for his school, but tried all other sports available. Through an active involvement in the Scout movement, he made many friends, played tennis and became interested in roller-skating and soccer. It is hard to find a sport—enjoyed by the predominantly Anglo-Saxon society to which Les Clisby belonged—which he did *not* play: football, cricket, swimming, hockey, tennis, squash, boxing and, when overseas, golf and skiing. He also was active with a local musical acting society, singing as well as acting in comedy sketches. It almost follows naturally from this wide interest in sports and group activites that Clisby was able to make friends easily. He created lasting bonds with people from all the groups of which he was a participant.

The Clisby family consisted of the parents, boys Max and Leslie, and girls Fay and Mary. The family home was 'Rosedale', named for the hundreds of roses which grew there. The house was a large freestone building in five acres (2 hectares) of gardens which provided flowers, fruit, vegetables and space for a cow to graze and some chickens to roam. The family was nearly self-sufficient. Later, the Clisby boys made a tennis court, and play was enjoyed by all the children and their friends.

Les Clisby came from a Methodist family, and on one occasion he was awarded a prize for bringing the greatest number of children along to the church services. He was a very caring and supportive young man, and would often come by 'Rosedale' to collect his younger sister, taking her to church.

Mr and Mrs Clisby encouraged the love of music in their children. Mr Clisby had a fine tenor voice and Mrs Clisby played the piano, providing the basis for many evenings of singing by the family and their friends. 'Wonderful pre-TV and radio days!' recalled Mary Clisby.

Somewhere, Les found time to indulge in a favourite hobby—buying old motor-bikes, stripping and rebuilding them, adjusting and tuning to perfection. Mary was allowed to assist by polishing the nuts and bolts, and rewarded by being given first ride on each newly rebuilt machine. She was devoted to her brother, as Les never made her feel she was in the way of such pursuits. Clisby eventually found employment, despite the Depression, but was underpaid and not happy with the firm. He also attended engineering classes at the School of Mines in the evening. It was this interest in engineering which led to his application to join the RAAF as a groundcrew member in 1935.

Newspaper and magazine articles published later, when Clisby as a fighter pilot was subject to some media attention, produced the standard line that he spent all his time as a boy and youth hanging around Parafield airfield, but this was not so.

After some time as an airman member of the RAAF at Point Cook, Clisby applied for a position on the next officer cadet course, and was accepted. He had been called into the office of the Chief Instructor, who suggested the application. After thinking about it for a few days, Clisby did so. His family in Adelaide were impressed and disconcerted by the arrival of a policeman, tasked to interview them and ascertain whether or not Leslie Redford Clisby was a fit and proper person to hold the King's Commission. In retrospect, how simple seem the vetting procedures of the mid-thirties.

Elsewhere in the world, Italy had invaded Abyssinia (Ethiopia) on 3 October 1935, and ignored sanctions imposed by the League of Nations. Earlier in 1935, the Saar region was returned to Germany, and a week later Germany repudiated the disarmament clauses in the Versailles Treaty, then reintroduced conscription. Hitler already had dictatorial powers under the Enabling Act passed after the burning of the Reichstag, and in following elections the Nazi Party had won government. He was appointed Supreme Commander of the German Army. Meanwhile, the Japanese had been waging war in China.

Cadet Clisby's nerves were tested on 24 April 1936. He had about 30 hours solo flying time, and was flying in Gypsy Moth A7-45 in formation with another Gypsy Moth near

A7-45 after Leslie Clisby abandoned it near Werribee, Victoria, on 24 April 1936. (*The Herald & Weekly Times Limited*)

Leslie Clisby after parachuting from A7-45 on 24 April 1936. (*The Herald & Weekly Times Limited*)

Werribee, Victoria. The aircraft were at 4,000 feet, about 100 feet (33m) apart, when

> suddenly the machine started on a right-handed turn of its own. I tried to correct this, and it was very slow in answering the controls, when suddenly it turned to the left. It started gradually to dive to the ground—I was unable to control it. The angle became steeper and steeper, the speed also. I tried everything I could to control it, but couldn't, so I next thought about getting out and using my parachute. The speed was getting faster and I stood up in the cockpit, then dived over. I don't remember pulling the ripcord until I saw the thing in my hand, and then I knew the 'chute was going to open. It did, with a jerk, and then I floated to the ground. I didn't feel nervous at all, it happened so quickly. I landed on soft dirt, went to a house, and rang up Point Cook.

Clisby had kept his nerve, and waited until he was well clear of the aircraft before pulling the ripcord, and then when he noticed that he was drifting towards some tall trees, pulled on the canopy lines to steer himself to open ground. He watched the Gypsy Moth dive to the ground and shatter to pieces. The engine was buried two feet (.6m) in the pasture.

In one of the coincidences of history, the place where Clisby landed was on a farm owned by a British migrant family named Lancaster. A son later joined the RAAF, was killed and is buried in France, at Choloy, near Toul. When Mr Lancaster ran up, Leslie's first remark was, 'What a sensation!'.

The newspaper accounts were quite detailed, and a little dramatic, one describing how the 'horrified eyewitnesses' of his 'leap for life' cheered as the parachute opened 'after hurtling groundward at terrific speed.' There were many references to Clisby's eligibility for the Caterpiller Club, comprised solely of those whose life had been saved by an Irvin parachute. At this time, he was only the second man in Australia to have done so, but there is no record in the Club's archives of his ever having applied.

Course 'A', July 1936 entry, RAAF Academy, Point Cook, Victoria. Leslie Clisby is fourth from right, front row, wearing corporal's rank. On extreme right is Bob Bungey, later to complete operational tours on fighters and bombers before his death in 1943. (RAAF Museum)

There was a great deal of investigation into the crash of the Gypsy Moth, and Clisby had to appear before two courts of inquiry, as well as undergo two medical tests. He wrote to his family, describing the events, and told how he was quite steady during the questioning. As far as he was concerned, it was very simple: 'the machine was doomed, the controls all went stiff. The machine wouldn't answer to them. I lost 2,000 feet, then dived out. I was too busy to get excited. I had every trust in the parachute. And to that I owe my life, through God. Don't worry please. Think of me as one who can look after himself now.'

However, in its thorough fashion, the Court of Inquiry interviewed every person who saw the crash, the instructors and cadets who had flown the aircraft, the groundcrew who serviced it, checked the maintenance history of the plane, and the wreckage was inspected by the very experienced Squadron Leader E.C. Wackett, Director of Technical Services, RAAF Headquarters. The Court found that there seemed to be no mechanical or structural failure and that Clisby's parachute jump was unnecessary, made because the aircraft flew into the slipstream of the leading Moth and Clisby did not realise the effect this would have, so thought the plane as going out of control. Only eight days previous, Cadet Chaplin was killed when the wings of his Moth folded vertically but he jumped too late for his parachute to open.

By this time, Les Clisby had determined to make flying his career, and the first step was to achieve a permanent commission in the RAAF by graduating from Point Cook. The curriculum included 22 subjects, and the habits and techniques of study acquired during his school days and while attending night classes at the Adelaide School of Mines now greatly assisted him. He did graduate, with the Passing-out Parade held on his 23rd birthday, 29 June 1937. In July, with 24 others, he left for England and a short service commission in the RAF.

The Dominions and Colonies were still very much part of the Empire/Commonwealth, and it was to the United Kingdom that young officers were sent for experience. Indeed, the Royal Australian Navy was part of the Pacific Squadron of the Royal Navy. Australia had no modern air force, though new aircraft were on order, so the RAF was used as a training organisation where the recent graduates could be given experience in flying and participating in a larger air force which had responsibilities around the world. On arrival in England, the 25 Point Cook graduates received commissions as Pilot Officers in the RAF.

By mid-1937, the international situation had steadily worsened. In 1936, Japanese militarists had appointed the Premier of Japan; German troops had entered the Rhineland, which had been occupied by the French at the end of the First World War; the Nazis received 99% of the vote in German elections; Abyssinia was formally annexed to Italy; the Spanish Civil War had broken out, with Italy and Germany supporting the anti-communist General Franco; two-year military service was introduced in Germany; Mussolini and Hitler signed the Berlin-Rome Axis agreement; King George VI came to the throne. In the first half of 1937, the war continued in Spain; the town of Guernica was heavily bombed by Franco's aircraft; US President Franklin D. Roosevelt signed the US Neutrality Act; Neville Chamberlain became British Prime Minister; Stalin began 'purging' the Soviet generals; the Japanese and Chinese continued fighting, but the Japanese soon were to occupy Beijing (Peking). When the young Australian officers were en route to England, at one stop they were asked to join one of the combatant sides in the Spanish Civil War, as mercenaries, but declined.

Leslie R. Clisby DFC

Clisby was one of six from the original group of 25 who were posted to 1 Flying Training School at Leuchars, Fife, Scotland for a refresher course, and to 'qualify' to RAF standards. He wrote to his family on 7 September 1937 telling them that the school was for training seconded Navy and Army officers, and that RAF pilots were first taught at a civil school, then sent to Leuchars with very little experience, for instruction on Service aircraft. 'Our flying is streets ahead of theirs — they think we are crack pilots. From here they go on to the Fleet Air Arm on fleet carrier ships, doing deck landings. I wouldn't really mind; I like ship life.'

The natural beauty of Scotland captivated the young Les Clisby, who had grown up in the very different southern area of Australia. He wrote home describing the scenery, and said he would simply stand and look out over the views, wishing the rest of the family could be with him to share the experience. He bought a car, and travelled as much as possible, writing full descriptions of the journeys for the family in Adelaide.

As well as flying and driving, Clisby also was enjoying hockey, soccer, golf and boxing at Leuchars. All this meant he had a full social life and when the course at Leuchars was completed in December, he had many invitations to visit the homes of the Navy officers.

He was posted to the leading fighter squadron in the RAF: 1 Fighter Squadron at Tangmere. 1 Squadron shared the base with 43 Squadron and recently 217 Squadron, a Coastal Reconnaissance unit, also had moved in. Les wrote that he was 'in the crack squadron of the RAF, so I am sure it will be good. I am flying Furies, single-seater aircraft, and we are being equipped with Hawker Hurricanes and Supermarine Spitfires some time this month.'

Clisby was the only officer on the station in the distinctive dark blue RAAF uniform. Presumably ever-mindful of economy, the RAF had agreed that RAAF officers could wear their original RAAF uniform until it wore out, after which RAF uniform had to be worn. Other Australians he met had by now purchased their RAF uniforms. Charlie Douglas from Adelaide was serving with Clisby, and while receiving instruction on the Hurricane at Brooklands motor racetrack, Clisby met Dick Reynell, from Reynella in South Australia. Reynell was a test pilot with Hawker.

The RAF, as it was to do so notably during the war, reflected the many countries which made up the British Commonwealth. At Tangmere, Les Clisby met and became friends with Hilly Brown from Canada, Caesar Hull from South Africa, Peter ('Prosser') Hanks and Peter ('Johnny') Walker from England. At Kenley RAF Station, he met fellow Australians Harry Hunter and Bill Fowler.

The pre-war RAF officers belonged to what was certainly the greatest flying club the world has ever seen. The over-regulation of post-1945 was not imagined, and aircrews enjoyed a busy social life when not flying. Les Clisby's letters to Adelaide described both aspects of his new life, from the Annual Dance at Tangmere, attended by the Duke and Duchess of Norfolk, to the 'tactical exercises and formation flying in 350mph (500kmph) Hurricanes and our old Furies which only do 170 flat out.' The Aussies regularly visited one another, as on the occasion when Les went to Headquarters 11 Group at Bentley Priory to collect some official documents. He stopped for lunch at Northolt aerodrome, and met fellow-Australians Skinner, Fowler, Leighton and five other new officers just arrived from Australia. Over Easter 1938, Clisby went with Johnnie Walker and Caesar Hull to Switzerland, where Walker renewed an acqaintance with the Dutch Consul, whose daughter and son hosted the pilots, and who placed his car and chauffeur at their disposal.

28 May was Empire Air Day, and a crowd of about 12,000

attended the show at Tangmere. Charlie Douglas and Les took part in the flying, and the squadron performed formation aerobatics, reportedly thrilling the spectators. All this was followed by what Clisby described as 'a huge party.' Two days later, the Inter-Service Sports were held, with Tangmere representing Fighter Command and Pilot Officer Clisby captaining the bayonet team. They reached the semi-finals, but were defeated by the Army. A lavish lunch was provided at the Officers' Mess, and Clisby commented on the 'yards and yards of gold braid' adorning the Army and Navy officers. At 'the show' in the evening, Clisby was in the box next to the Duke of Kent and Princess Marina, who he thought 'looks rather nice'.

Next day, Clisby went with Hilly Brown to Cracknor Hard, where the 50 ton Brixsome trawler *Devon's Pride* was moored. The boat was owned by Pilot Officer Pennington Leigh's father, who had made it available to the pilots for a 2½ week cruise. Mr Pennington Leigh skippered the boat, while aboard were Hilly Brown, Caesar Hull, Ross Turner, Killy Kilmartin, Charlie Douglas, Chuck Mayo, Wing Commander Hill, Clisby and young Pennington Leigh. In all, there were four Aussies aboard, and Kilmartin had lived in Australia for many years. The boat had enough food for a month, and sailed out for a leisurely cruise along the coasts of France and southern England. Pennington Leigh, Hull and Kilmartin were from 43 Squadron, and had to be back earlier than the others, so they were simply rowed ashore at Littlestone on Sea and left there. However, not having shaved for two weeks and wearing scruffy sea-going clothes, the trio were not the accepted image of RAF officers in 1938 and were believed to be aliens, being held in custody for five hours until RAF Tangmere vouched for them. Meanwhile, the rest had sailed merrily away, unaware that HM Customs had been chasing *Devon's Pride* since the beginning of the erratic holiday. Next day the 43 Squadron pilots found the boat and proceeded to fly low at and around it, which Clisby admitted 'absolutely scared me stiff'. The day after, Pennington Leigh brought the boat back to his permanent mooring at Poole, the holidaymakers went ashore, and back to Tangmere, after what Les said was 'the most exciting and thrilling time of my life.' It was a long, long way from rebuilding old motorbikes in suburban Adelaide.

The past year had seen a further deterioration in the world situation. Restrained only by the statesmanly protests of the weak and irresolute democracies, the dictators expanded their territories as they saw fit. In the second half of 1937, Germany, Italy and Japan had strengthened their alliance; Japan continued to capture cities in China; Hitler ignored requests to leave the matter of the Sudeten Germans. In 1938, he became supreme commander of all German military forces; in March, Austria was occupied by the German armed forces and became part of Germany; Hitler announced his intention to destroy Czechoslovakia; the Japanese and Russians fought on the Manchukuo border. A most significant factor in all this was that Hitler required all German servicemen, regardless of rank, to swear an oath of loyalty to him personally, not to the state or government. Along with all the other pressures on them in time of war, this personal oath of loyalty was regarded by many German officers as inviolate, and used to advantage by the unscrupulous Hitler.

In August 1938, 1 Fighter Squadron attended an Armament Training Camp at Sutton Bridge. Leslie Clisby was appointed squadron Armament Officer, with two sergeants and 15 airmen 'to look after and keep working.' In reality, of course, Pilot Officer Clisby would have little to show the sergeants and airmen, but presumably he was astute enough not to get in their way. Les was mainly occupied with the squadron commander, Squadron Leader Bertram, making up new programmes and compiling returns after the day's firing. Bill Fowler RAAF, with 3 Fighter Squadron at Kenley, also was present, and the two officers in dark blue were called 'The Diggers'. It seemed war was imminent, and the bright silver colour of the RAF fighters was quickly sprayed and brushed over with drab green and brown.

In Europe, Prime Minister Chamberlain was trying to deal with Hitler over the problem of Czechoslovakia and, on 30 September, Britain, France, Italy and Germany agreed to transfer the Sudetenland to Germany. Chamberlain returned to England and at the steps of his aircraft held up the paper signed by Hitler, which was to bring 'peace for our time.' Encouraged by this dismemberment of a neighbour, after the Germans took the Sudetenland, Poland took Teschen, Italy 'readjusted its borders' with Hungary and Czechoslovakia whilst Hungary annexed Southern Slovakia. In China, the Japanese took more cities; in Germany more anti-Semitic pogroms were launched.

Just after the Munich Crisis, Clisby went into hospital to have his tonsils removed. Unaware of the true state of Britain's defence forces, Les was disappointed in Prime Minister Chamberlain's giving in to Hitler's demands by 'running to him, begging for peace, and then giving him everything he asked for in the first place. All very well for peace, but what about England's prestige? It has certainly had a jolly good kick in the pants. What with the finest navy and air force in the world, I don't see how we could go wrong.' Clisby described the preparations for war in Britain, and the issue of millions of gas-masks, which made him very worried for the future of England 'if it doesn't do something very quickly.' He thought that there would be four or six months peace, while Hitler digested his latest gains, and then demands for the Tyrol would be made.

By 15 November, Clisby was back on flying duty after spending the last days of sick leave on *Devon's Pride*. He was still adjusting to the high speed of the Hurricane, and the distance which could be covered at 300mph (480kmph). He wrote that the fighter could 'land just like a bird, and have brakes almost as powerful as a car. One can land as slow as 80mph (128kmph), and still there is plenty of control.'

On 12 December, Les informed his family that he had decided to become engaged to a girl he had met in Melbourne in 1935, who had spent holidays with the family in Adelaide. The couple had been exchanging letters since he sailed from Australia.

Clisby enjoyed a traditional Christmas with the family of Squadron Leader Bertram, and described the roaring fires, holly and mistletoe, candlelit dinners, the company of Hilly, Duggie and Caesar, with the invited daughters 'of Colonels and retired sea captains. Had the best Christmas one could possibly have had, thanks to the Bertrams.'

It was back to Tangmere, flying and sport in the New Year of 1939. On 25 February, Les broke his collarbone and dislocated his shoulder at rugger, resulting in more sick leave, time on *Devon's Pride* and with the Bertrams. The first personal tragedy in Les' RAF service came on 26 March, when Charlie Douglas was killed flying in thick fog. Les was so upset that he could not write home about the matter, except to say he was on the committee of adjustment, which sorted personal belongings and returned them as appropriate to the RAF or family.

In April, there was another Armament Training School at Sutton Bridge, and here Clisby met 'the Father of the RAAF', Air Vice Marshal Richard Williams, who was happy to see another dark blue RAAF uniform. Les also met the man who was to be his leader in combat—Squadron Leader Patrick 'Bull' Halahan. At this time, he described Halahan as 'a good chap, very keen, and full of heartiness.'

Halahan made a significant alteration to the squadron gunnery arrangements. At the time, along with the stylised 'Fighter Attacks' developed over the decades of peace, the armament of the new eight-gun fighters was sighted so that a target the size of a bomber would be covered by a spread of shot from the guns. Halahan reasoned that concentrating the shot from all eight guns onto one spot, at 250 yards (metres) range, would give far more lethal results than a spread pattern. He proved it on the towed targets during this camp, and on his own authority had all squadron aircraft altered accordingly. This was against regulations and he could have been court-martialled. At the time, it was quite reasonably assumed that the only German aircraft which might be met over England during a war would be bombers, as fighters simply did not have the range to escort bombers from bases in Germany. No one thought that Luftwaffe fighter units would be based just across the English Channel.

Elsewhere, the war in Spain continued, with the fascist forces gradually winning; German troops marched into Prague and occupied Bohemia and Moravia. Hungary seized the opportunity to annex Ruthenia. Chamberlain accused Hitler of breaking his word, and five days later Germany annexed the Memel, forcing the Lithuanians to sign the relevant treaty; Poland rejected Berlin's proposals to join the German city of Danzig (Gdansk) with Germany; Chamberlain announced Anglo-French guarantees to Poland. Conscription began in the UK; Germany signed a military alliance pact with Italy, then a non-aggression pact with the USSR. Chamberlain warned Hitler that the guarantees to Poland would be fulfilled. Hitler was intent on more territorial demands. As the Polish crisis deepened, and war preparations intensified in Britain, Les wrote that 'the European situation certainly is a little precarious—in fact it is rather grave, but I don't think you in Australia have anything to worry about. All my business is in capable hands. I am perfectly happy in doing what I do, and am pleased to say I'm quite proficient at my job, which is much better than having to do something which I know nothing of.'

By the time the letter arrived in Adelaide, the British Commonwealth, with Australia, was at war with Germany. Gone forever were the days of the carefree pilot enjoying sailing holidays, rubbing shoulders with British nobility, dropping in for lunch at another airfield, thrilling the crowds with aerobatics.

Since 1936, Britain and France had been meeting to discuss common matters of military importance in Europe, and some agreement had been reached on the size and dispositions of British forces on the Continent. The Advanced Air Striking Force (AASF), of 10 bomber squadrons, was to be accompanied by what was called the Air Component of the British Expeditionary Force (BEF). The signing of the non-aggression pact between the Nazis and Soviets on 21 August was an obvious sign that the dictators had come to an agreement and that Germany would be free to launch military operations against Poland. Three days later, on 24 August, Britain secretly mobilised the RAF elements to be sent to France. On 1 September, Headquarters of the AASF opened at Reims. Next day, 10 squadrons of Fairey Battle light bombers arrived.

War did not bring the expected swarms of German bombers raining destruction as expected in modern war. The Italians over Abyssinia, the Japanese over China and the Fascists over Spain had given stark examples of the impersonal death and devastation which was to be expected, but nothing happened over Britain. Germany left a force to guard her western borders and got on with the task of obliterating Poland while perfecting the 'blitzkrieg' techniques. Four squadrons of Hurricanes were to go to France as part of the Air Component of the BEF, but Air Chief Marshal Hugh Dowding, commanding Fighter Command, protested over sending *any* force. His Fighter Command had not yet reached the agreed minimum for defence of the UK, let alone send squadrons to France and so result in losses which also had to be made good. Fighter Command had 57 squadrons, of which 46 were for the Air Defence of Great Britain (ADGB), four were for the Field Force — the British Expeditionary Force (BEF) — and seven others were for separate tasks. After inspecting the squadron, Dowding told Squadron Leader Halahan of his reluctance to release units for France, adding that only 12 fighters and the pilots present that day would be sent — no reinforcements could be spared.

The RAF force for France, as noted, was in two parts. Air Marshal A.S. Barratt commanded the British Air Force in France (BAFF), in which Air Vice Marshal C.H. Blount commanded the RAF Component in the BEF, of four fighter squadrons, four Army co-operation squadrons, and four reconnaissance, close support and fighter protection squadrons, and Air Vice Marshal H.P. Playfair commanded 10 squadrons of Fairey Battle bombers in the AASF.

Les Clisby wrote to his family, reassuring them and urging that they not take too much notice of the newspapers. He mentioned the 'absolutely perfect' weather the squadron was enjoying. 'I'm looking brown and healthy, and so are all the boys.' He said that the squadron worked from dawn to dark, then stood by for whatever might happen at night.

When hostilities began, Reservists reported to their allocated unit, and 1 Squadron received about 100 of these men. All fitted in well, and the new transport drivers showed their mettle when they were sent off to collect the extra vehicles — they returned with several more than they should have taken, including a mobile workshop, to which the squadron was not entitled. Halahan was quite pleased and realised that the new men were fast becoming part of the squadron family. The entire squadron enjoyed a definite 'family' feeling. Halahan believed, and wrote in the squadron diary, that he could not have a better squadron to take to war.

Clisby, with the other officers, had been sobered by a lecture given by Squadron Leader Coop, former Assistant Air Attache in Berlin. Coop provided them with the facts about the Luftwaffe, its size and quality. Little of this information had been available before war was imminent, and the average squadron pilot had no clear idea of the enemy air potential.

On Friday, 8 September, 1 Squadron taxied out behind Squadron Leader Bull Halahan and took off for Le Havre. As the first British fighter squadron to land in France, and the premier squadron of the RAF, the Hurricanes did a fly-past for the entertainment of the locals, then went in to land. The evening was spent enjoying the attractions of the town, helped along by a large number of American citizens who were leaving the war zone.

The morning was a time of regret. Halahan decided air raid trenches were to be dug in the nearby orchard, and all ranks sweated and suffered alcoholic remorse while digging for three hours. After recovering with the aid of buckets of well water, the pilots went to their aircraft. Halahan intended to do a little formation flying to demonstrate just how good 1 Squadron was, even after a night on the town.

Halahan led, with Hilly Brown and Les Clisby to port and starboard, and the remainder of the squadron in flights of three. They took off, concentrating on maintaining formation while retracting undercarriages, and Halahan proceeded to put them through the pre-war standard attacks for use against bombers, formation and individual aerobatics, and a spectacular low flight at tree-top level before landing. It was time for lunch, for those with the strongest stomachs, and then more trench-digging.

73 Squadron arrived in the afternoon, the second fighter squadron to come in; 85 and 87 Squadrons were to follow. Later, 1 Squadron found that the 73 Squadron pilots had gone into town and reaped the rewards of the flying display by 1 Squadron earlier in the day. A disgruntled 1 Squadron had left for Cherbourg, weary after more digging.

The airfield at Cherbourg was rather small, requiring some good landing techniques, but all the Hurricanes got down safely. By this time, the squadron was a group of tired young men who gratefully fell into bed after dinner.

Next day, Sunday, was one of low cloud, with bases at 200 feet, or even 50 feet in places, but a patrol was flown over the ships from England arriving in the port, bringing the leading elements of the British Expeditionary Force (BEF). Next day the squadron returned to Le Havre, and remained there for 17 days of inactivity. On 29 September the squadron flew to an airfield at Norrent Fontes, near St Omer, and settled in. First of all, latrines had to be dug and, to improve tempers, it began to rain.

Clisby wrote home again, reassuring his family. 'I've no fear for anything I might do. All I can say is, I'm very sorry for the German people. It does seem wicked to think all this can be caused by a few rats like Hitler, Ribbentrop, Goebbels and Goering.' The current Allied propaganda was that the German people did not support Hitler, but in reality they were to do so until the bitter end. Another facet of the propaganda effort was that Germany could not wage a war anyway, as the economy was about to collapse, and Les added, 'I'm positively certain that the German forces will collapse in time, due to economic reasons.'

The same wishful thinking was applied to the Japanese, who supposedly were on the verge of collapse after fighting in China. Actually German production was to rise throughout the war, except for the last few months, and the Japanese were defeated only after their cities had been razed and they hardly had a ship afloat.

The keen fighter pilots were not too pleased to learn that the first German aircraft shot down by the RAF had fallen to the rear gunner of a Fairey Battle, when a Messerschmitt Bf109 was destroyed on 20 September. On 30 September, the qualities of the opposing forces were demonstrated in significant actions. A formation of three 88 Squadron Battles was attacked by Messerschmitt Bf109s, after the French anti-aircraft guns had hammered the RAF aircraft. One 109 was shot down but two Battles were lost. Then five 150 Squadron Battles were attacked by Messerschmitt Bf109s, and the British pilots and gunners did as they had been trained — maintained formation and tried to bring the attacking fighters into the cross-fire of the gunners. Four Battles were shot down, as the slow under-armed bombers, with single .303-inch machineguns protecting the rear, were no match for the fast cannon- and machinegun-armed Messerschmitts. The formation claimed four 109s as well, but the remaining Battle had to crash-land at base and was written off. The French fighter escorts had not been making rendezvous, and after this debacle informed the RAF that escorts would no longer be provided anyway.

Another result of the combats was that it was decided the Battles would go in to attack at low-level rather than suffer fighter attacks at higher altitude. This was a case of out of the frying pan and into the fire, though no one realised it yet, as the German flak defences had not been experienced.

Squadron Leader Halahan went to the Reims Headquarters of the AASF, and offered the Hurricanes as fighter escort.

This was agreed on 10 October, and 1 and 73 Squadrons were moved closer to the frontline, with a portion of the line to patrol. 1 Squadron was allocated the right sector, Nancy-Metz, and 73 was given the left sector, Thionville-Verdun; either could be assigned the centre or elsewhere. French squadrons were responsible for areas to north and south. To replace the two squadrons in the Air Component, 607 and 615 Squadrons, equipped with Gloster Gladiator bi-planes, were sent from England and tasked with airfield defence duties. 1 and 73 Squadrons moved to Vassincourt airfield, near Bar-le-Duc, 50 miles (80km) east of Reims. Squadron morale rose considerably, and there was some pity for 85 and 87 Squadrons, held back for convoy patrols over the Channel. Halahan led the first patrol which could be termed as operational, an hour-long sweep over the lines to Saarbrucken, through flak, but with no sighting of enemy aircraft.

The rain continued for two weeks. There was little flying, and no contact with the enemy. The period of the 'Phoney War' had begun. The Poles had been annihilated. It was anticipated by clearer minds among the Allied forces that Germany would attack in the West, and would violate the neutrality of Holland and Belgium. Hitler had agreed to observe the neutral status of both, but he had not kept any promise, so only the naive expected him to do so in this case. The French forces were supposed to be modern and capable of defending their homeland, but little offensive action was expected. Vassincourt, 1 Squadron's new base, was on a hill, with woods around it, near the village of Neuville. The squadron was billetted in the village, and immediately struck up friendships. The Germans had been in the area in the battles of 1870 and 1914, so the locals were not inclined to welcome them again. Being closer to the front, there were sightings and a few combats. The first victory was scored by 'Boy' Mould on 1 November, when he took off after a high-

RAF Hurricanes over France. (Mary Eckert)

flying Dornier reconnaissance aircraft, came in astern and fired a long burst, setting the German on fire. The Dornier went into a spiral dive, exploded in a field, and there were no survivors. A machinegun and holed oxygen bottle were taken to the squadron Mess as trophies of the first British fighter victory of the war.

This high point was followed by three weeks of bad weather, but 24 November dawned bright and clear. The Luftwaffe seemed to be keen after such a break, and there were several combats with reconnaissance aircraft, the Hurricane squadrons claiming a total of five Dorniers and a Heinkel, the latter shared with the French AA. When attacking the Heinkel near Metz, Sergeant Clowes lost part of his rudder after being rammed by an enthusiastic Frenchman in a

Morane, but he landed at Vassincourt. A Flight destroyed a second Dornier, but as the flight leader, Pussy Palmer, was flying close to check it, the pilot suddenly throttled back, slid behind the Hurricane and hit it with the single forward-firing machinegun. Palmer force-landed with his wheels up, lucky to be alive after a bullet came through the back of his seat, narrowly missing his head.

The brave Dornier pilot also force-landed and was then entertained to dinner by the RAF fighter pilots. However, as he was the property of the French, it took some persuasion before this visit could be accomplished. The pilot, Unteroffizier Arno Frankenberger, made a good impression on the squadron, as a fellow airman, and Halahan thought that the visit had been a mistake, as his fighter pilots saw the human face of another pilot, not an enemy fighting for an evil regime.

Frankenberger told them that the Luftwaffe went on leave on the weekends. Halahan carefully checked this allegation of greatly reduced flying with the aid of a newly arrived radar set, and discreet reconnaissance flights, then convinced his higher headquarters that the RAF fighter pilots also should be allowed, within reason, a turn at weekend leave.

Of more importance, the combat resulted in rear armour being fitted to the fighters. Considered opinion in the RAF hierarchy was that installation of armour behind the pilot's seat would affect adversely the centre of gravity, so the request was denied, but Squadron Leader Halahan went ahead and acquired some from a wrecked RAF Fairey Battle light bomber, fitted and test-flew it in a Hurricane, then told Air Ministry what had been done. He sent Hilly Brown to England to demonstrate the modified Hurricane to the non-believers, and all RAF fighters received armour protection for the pilots.

In November, the AASF flew 237 patrols, involving a total

of 369 sorties, while the Air Component flew 163 patrols, totalling 512 sorties. Again things became quiet. For six weeks, there was no combat, and winter fastened its grip on Europe.

On 20 January, the RAF fighters in France totalled six squadrons, organised as: 67 Wing, of 1 and 73 Squadrons; 60 Wing, of 85 and 87 Squadrons; 61 Wing, of 607 and 615 Squadrons.

Les Clisby wrote home of a pleasant surprise he had enjoyed. The squadron was visited by Mr Fairbairn, Australian Minister for Air, his secretary, and Group Captain Mac-Namara VC RAAF. The party stayed for lunch, and after a long conversation, MacNamara told Les to call on him when next in London, and Mr Fairbairn told him that when the first RAAF fighter squadron was formed in England, Les would be posted to it as a flight commander. Newsreel film was taken of the Australians walking and talking, so Les warned the family to look for the sequence in the theatres. His own letters home reflected the homesickness he seemed to be feeling, perhaps accentuated by the combination of lack of mail and the miserable weather, which meant there was little opportunity for action.

At the end of January, Les went to London on leave, called on MacNamara, and also spent time with the families of Squadron Leader Bertram and Johnny Walker. He was made welcome, and wrote home that he had 'been to another home in England and found some people after my own family's heart.'

January 1940 was freezing as Europe suffered the worst winter since 1918. Established as it was 'in the field', the squadrons of the RAF contingent endured bleak living and working conditions, and the groundcrews persevered with their daily tasks in inadequate canvas hangers. February was quiet, and BAFF managed only 25 sorties over four days in the month. On 2 March the squadron claimed its fifth victory, when Hilly Brown, Soper and Mitchell, a New Zealander, attacked a Dornier Do17 reconnaissance plane. Brown's propellor disintegrated, and he went down to land, Soper fired, but the German gunners hit Mitchell in the engine, and while trying to land he crashed and was killed. The Dornier went down, on fire. Next day, Brown and Soper attacked a Heinkel He111 bomber, sending it down to crash-land between the lines of the opposing armies, and two of the crew were seen to escape to their own side. Les described these victories briefly in letters, but told his family there had been no squadron losses.

Intensely cold bad weather halted operational flying. It was impossible to keep warm inside the buildings, and the pilots and groundcrews who had to stand readiness in tents at the airfield suffered greatly. The squadron had not engaged German fighters so far, though by the fortunes of war 73 Squadron had fought them several times. The Messerschmitt Bf109 (often named ME109) had been the subject of a great deal of propaganda from the Germans, and the aggressive and competitive RAF pilots were keen to know just how good the Messerschmitt was. Cobber Kain, from New Zealand, and Fanny Orton, both 73 Squadron, had both been successful over Messerschmitts, but in his first five combats Kain had been shot down twice.

The Luftwaffe propaganda machine also spent a great amount of effort on the Messerschmitt Bf110, 'Zerstorer' or 'Destroyer', which was reputed to be a formidable military aircraft, but none had been met in combat. In what could be a move typical only of the British, the prize for destruction of the first Messerschmitt Bf110 was to be, not a medal, but dinner at the expense of Air Marshal Barratt, the Air Officer Commanding (AOC). In 1 Squadron, Bull Halahan told his pilots not to be foolish and chase Germans back over their own lines, but to be patient — there would be enough fighting when the German land offensive began. Until then, such enemy as appeared would have to suffice.

As the weather cleared and the first signs of spring appeared, there was more flying and more opportunity for combat. As much time as possible was devoted to training new pilots who arrived from the British training system. These men had not the benefit of a lengthy pre-war training cycle. Some had only a dozen or so hours on Hurricanes, no formation flying practice, no oxygen experience, no gunnery training, no radio-telephone training and no 'fighter attack' training. A similar situation existed in the bomber squadrons.

The sunshine was probably welcomed by no one in 1 Squadron more than Les Clisby. Adelaide winters simply were not as harsh as those in Europe, and he wrote on 27 March, 'Thank goodness the weather has taken a turn for the better. Real Australian spring weather, quite warm flying at 25,000 feet — comparatively, that is. Hope you heard the news re the Western Front aerial battles yesterday — we caught a horde napping. It's a pity I can't tell you all the details — censor you know! For us the war is at last starting, thank goodness; getting a little fed up. I have again been acting as flight commander, quite a responsibility, but keeps one occupied and busy.'

In the battles 'yesterday', he probably was referring to a combat between 73 Squadron and a formation of Messerschmitt Bf110s. The Hurricane pilots were pleased to find that their fighter was able to out-manoeuvre the 110 even though the Messerschmitt was faster and could climb higher. However, none were claimed destroyed. Cobber Kain, the New Zealander, claimed two 109s destroyed but was shot down himself, and in a series of other combats with Dorniers and Messerschmitts, a total of five were claimed destroyed.

The dinner given for destruction of a Bf110 was awarded

to three 1 Squadron pilots, who shared in the victory on 29 March — Flight Lieutenant Johnnie Walker, Flying Officer Bill Stratton and Flight Sergeant Arthur Clowes. Air Marshal Barratt took them to Maxim's, and it was said that the collection of widely differing ranks at his table created amazed looks on the faces of other diners.

On 31 March, between 13.30 and 18.30, in Hurricane L1856, Clisby flew three patrols with no contact. He had flown many others since arriving in France on 8 September. In March, BAFF had flown 310 sorties.

At 11.25 on 1 April, again in L1856, Les took off as Red 3, in a section led by Prosser Hanks, with Boy Mould as Red 2. Hanks led them to 22,000 feet at Thionville-Boulay, and they had only been patrolling for five or six minutes when AA fire drew their attention, but there seemed to be no aircraft near the bursts. Clisby continued searching the sky and in another direction saw nine Messerschmitt Bf110s at 26,000 feet, flying west. He informed Hanks, who began climbing onto the tails of the Messerschmitts but the 110s saw them and six dived to engage. Hanks fired head-on at the leading 110, hit it, but received a cannon shell in his port wing, and the explosion holed his petrol tank, oil tank, main spar of the wing and his perspex canopy, but he was able to continue the combat, shooting at four separate Messerschmitts, before going on to land safely at Rouvres.

Les had slid into line astern, and when he saw Hanks and the 110 firing at each other, moved to the left and engaged another Messerschmitt, which went through a series of steep dives, climbs, stall turns and steep turns, with Les hanging on and firing. He used all his ammunition in six bursts, and saw flames and smoke coming from the fuselage and port engine of the 110, which went into an almost vertical dive from 18,000 feet. Clisby followed, but lost sight of it 'somewhere north of a line between Thionville and Longuyon.' He landed at Rouvres to refuel and re-arm, then returned to Vassincourt. The 110 was obviously badly hit, but when his combat report was written, no wreckage could be found in Allied territory.

All three claimed a 110 destroyed, but no confirmation of the victories was possible when the squadron records were reconstituted in England after Dunkirk. However, in his history of the squadron, Michael Shaw states that three 110s were confirmed by wreckage on the ground. Luftwaffe records show that nine 110s of I/ZG26 were escorting a Dornier Do17 of 3(F)/11; in the combat one was damaged, and although the pilot, Oberfeldwebel Blass, was wounded, he reached base.

Flying Officers Lorimer and Clisby, France, 1940. (Mary Eckert)

It seems from Clisby's combat report that only his port guns fired, and of those four machineguns two had stoppages, and only two fired the full amount of 300 rounds, with a total of 990 rounds expended in the six bursts.[1]

Next day, 2 April, Les flew the early morning patrol at 06.25, with Hanks and Mould, but there was no contact. At 10.30, he flew in L1297 as Red Leader (A Flight), with Flying Officers Palmer, Kilmartin, Richey and Lorimer. Johnnie Walker was away, and Clisby was acting as leader. The weather was cloudy, and again AA indicated enemy, a twin-engined aircraft at 20,000 feet 15 miles (24km) south-west of Metz. As the Hurricanes closed on it, a trail of white smoke appeared behind the unidentified aircraft, and it then out-climbed and out-distanced the Hurricanes, turned east and left them. Post-war records show this to have been a Dornier Do17. The flight was then at 20,000 feet over Puttellange.

With their attention focussed on the Dornier, Clisby's flight did not search the sky thoroughly and failed to see a *schwarme* of four Messerschmitt Bf109Es above. This four-some from III/JG53 was led by the great Werner Molders, already an ace with 14 victories in the Spanish Civil War and seven in this war.

Palmer was 'tail-end Charlie' and resumed weaving to keep a lookout astern even during the pursuit, saw enemy above, and reported them. Clisby looked and identified two widely separated Messerschmitt Bf109s diving, but as he turned the flight to the right, Palmer was attacked by Molders, whose accurate burst of fire set alight the petrol tank; Palmer bailed out. He landed safely in Allied territory, but only 500 yards (metres) from the German lines. French soldiers helped him to safety and extended considerable hospitality before delivering him to Vassincourt.

Clisby saw the German fire, the smoke stream from Palmer's Hurricane and a black object — Palmer — fall away from the fighter. Molders broke left and down, and Clisby went after him, closed to 250 yards (metres) and with slight deflection, opened fire. The 109 began to straighten out but Clisby fired again and white smoke, then black, poured from the diving Messerschmitt, which rolled onto its back and went down vertically, leaving a long trail of black behind it, right into the clouds. Cloud tops were at 5,000 feet and at that height Clisby began to pull out of the dive, pulling the control column as hard as he could, winding back on the tail trim to assist. The centrifugal forces almost made him black out and he realised he was at only 500 feet, north of Puttellange, south of Saarbrucken. He turned south-east and flew in that direction until he recognised Nancy, then returned to Vassincourt. In this combat, all eight guns had fired, a total of 1682 rounds.

Killy Kilmartin and Clisby each claimed a 109, and Paul Richey made a forced landing. Kilmartin fired all his ammunition into the second 109, which began to burn, rolled and spun down. Kilmartin saw it crash into a French village. A

Werner Molders was engaged by Leslie Clisby but escaped and was the first German pilot to claim 100 victories.

total of nine Messerschmitts were believed involved in the combat. All four of Molder's 109s returned to base and Molders claimed his eighth victory since September 1939. These were early days in the war, and it was not realised that the Daimler-Benz DB601 engine emitted a thick trail of black smoke when suddenly pushed to full throttle. Many 109s were claimed by eager French and British pilots who last saw 'their' Messerschmitt going down streaming thick black smoke.

Molders had gone to Spain in April 1938, to the third fighter squadron of the Condor Legion, and soon after the unit received the new and fast Messerschmitt Bf109. Molders began to score and, like many other German pilots, acquired invaluable experience in flying and fighting modern combat aircraft before the Second World War began. He scored 14 victories in Spain, to become the leading German ace of that war. On this day, he was on his 71st combat sortie in the Second World War and had shot down five French and two British aircraft. After 101 victories, and removal from combat on the orders of Field Marshal Goering, Molders was killed in a flying accident on 22 November 1941 when returning for the funeral of General Ernst Udet.

The weather in France thickened to cloud and rain, which prevented flying. This continued for several days, and only training flights for new pilots were possible on 5 and 6 April. Some new aircraft also were delivered. This draining away of men and aircraft was what Air Marshal Dowding had tried to prevent, although the RAF squadrons were gaining experience in actual combat conditions. 1 Squadron was proving the superiority of the 'spot harmonisation' of the Hurricane guns ordered by Bull Halahan. The German aircraft, like the RAF, were lightly armoured, if at all, and the effect of this stream of bullets resulted in great damage and destruction, provided the pilot shot accurately enough to hit his target.

Hitler issued orders for the invasion of Norway and Denmark; an Allied force sailed to Norway but was narrowly beaten there by the German invasion from sea and air on 9 April.

On that day, Les wrote from the Charing Cross Hotel, Strand, London. He was on ten days' leave, and intended to spend some of it with Johnny Walker's family, 'a change after the hectic time on the Western Front. Damn good fun — we'll beat them I'm sure — they aren't made of the stuff we are.' He had managed to leave France on the day all leave for the British forces there was stopped. Before going to Walker's family, Les had called on Group Captain MacNamara at Australia House. He was somewhat embarrassed to find that it was intended to make him into an Australian hero. He was introduced to the Australian High Commissioner, Stanley Bruce, and had to recount his exploits in France. MacNamara told Bruce the details of Clisby's RAAF career, from enlistment as a mechanic, through graduation from Point Cook, to going to France with 1 Squadron. Newspaper reporters and photographers were so intrusive that MacNamara had to ask them to desist. Les was invited, and of course could not refuse, to lunch with MacNamara in a few days, so had to return to London for the occasion.

Meanwhile, in France, the squadron was warned for a possible move to Berry-au-Bac, but the information was not definite enough to begin packing. Of course, when the order was confirmed, it also required the move to be completed in very short time, and within two hours the aircraft had flown out and 45 minutes later were at readiness at Berry-au-Bac, with the equipment and personnel to follow. This was done by 03.00 on 12 April.

The squadron officers were accommodated in a chateau with the officers of 142 Bomber Squadron, with the men at

Leslie R. Clisby DFC

Flying Officer Leslie Clisby RAAF, on leave in London, April 1940, after several combats with Messerschmitts over France. He is not in dark blue RAAF uniform, nor does he have 'Australia' shoulder titles. (Mary Eckert)

uncomfortable billets in Ville-aix-Bois and Juvincourt. With little sleep, the men were at work before dawn, and the squadron records included praise for the way in which all performed so that by midday the maintenance and flight tents were erected and the unit was settled in — less than 24 hours after moving from the previous airfield. The pilots took over a windowless ammunition storage building, and made it livable with a few boxes, a stove and a gramophone. Cloudy weather and rain precluded any flying.

On 13 April weather again was so bad that no flying was possible. Next morning, 14 April, the only flight was an attempted reconnaissance by two Hurricanes, which were forced back by weather. Later in the day a strong wind swept away the armoury tent, then it rained, and all this 'added nothing to the comfort of our armourers', as the squadron records noted. On 15 April solid cloud and rain continued to keep the squadron grounded, and the day was spent trying to locate suitable accommodation for the entire unit closer to the airfield, and establishing a working telephone line from the billets to the aerodrome. The only outside line was to operations at Wing Headquarters, and all messages other than operations matters had to be taken by despatch rider.

Squadron Leader Halahan, after untiring effort, arranged for the squadron to be billeted in the village of Pontavert, which had only been built since the First World War and was a great improvement on the other, older, villages. This move was accomplished on 16 April. On 17 April, fighter affiliation exercises were flown with the Battle light bombers of 12 and 142 Squadrons, mainly to train newly arrived pilots. Halahan himself flew a reconnaissance along the borders near Belgium and Luxembourg. The squadron record commented that the pilots were fed up with the prolonged inactivity and the long useless hours at readiness. More affiliation exercises were flown on 18 April, but the main item of interest was delivery

Leslie R. Clisby DFC

of a new Hurricane with a constant-speed airscrew, which improved performance. It was flown by Halahan and other pilots, all of whom were pleased with it. Orders came to return to Vassincourt. 19 April was taken up with the return and in the evening the officers were guests at a 'welcome back' dinner at the Hotel de Metz.

20 April saw more combats. A Heinkel He111 escorted by three Messerschmitt Bf109s (reported as two 109s and a Heinkel 112) at 25,000 feet near Charnay was attacked by Flight Lieutenant Prosser Hanks, Boy Mould, Berry and Albonico. The Heinkel and two Messerschmitts were claimed destroyed and the other was last seen in a steep dive, pouring black smoke. At about the same time, Johnny Walker led Hilly Brown, Billy Drake and Stratton against nine Messerschmitts, claiming three destroyed and one 'out of control'. Walker's victory crashed in Allied territory, Brown's fell into Germany, and Drake chased his to ground level, into Germany, and saw it hit a hill. The chase had been over tree-tops and into valleys, and the 109 pilot flew straight for some powerlines, obviously intending that Drake crash into them, but Drake slipped underneath the cables and, as the 109 was cresting a hill, Drake was able to hit it with a telling burst. It caught fire and he watched it go, burning, into the hillside. At 09.45, Killy Kilmartin took off after a high-flying enemy, and eventually chased it to low-level, but was hit in the engine and forced to land. However, the German aircraft crashed further along, and was found to be one of the new Junkers Ju88s.

For this day, the squadron record stated that all the original squadron members had been in combats, but the unit had worked well together and 'made it a squadron "show" without any publicised individuality.' This reflected official RAF policy, which did not admit to the idea of the 'fighter ace' so sought after by the media, but insisted on credit shared equally for such successes as were gained. However, as in the First World War, journalists were to seek out victorious pilots and, despite censorship, create heroes for the public no matter what the wishes of RAF commanders.

The pilots were a little annoyed by the sensation-seeking journalists, especially as they knew little about the subject matter. Finally, Halahan banned reporters from the Officers' Mess. A few journalists did become accepted, particularly Noel Monks, who wrote in some detail about the fighter squadrons. His hotel room in Reims was a popular meeting place, and was more enjoyable for the opportunity to have a hot bath. The team spirit and family atmosphere which prevailed in 1 Squadron was mentioned by Noel Monks, who added that the reporters somehow felt like outsiders when with 1 Squadron, but at home with the individualists of 73.

It seems from Clisby's letters that 1 Squadron preferred the 'team' or 'family' image; there was some disdain for a pilot in 73 Squadron who was receiving a great amount of media attention, who had only two confirmed victories but was credited by the journalists with more after every combat. The pilot was described in one letter as 'such a fool' and in another the remark was made that there were many other worthy pilots in France as well as the one in question.

On 25 April, Les Clisby wrote again to his family. 'Well, here I am back after a perfect leave. I had the most marvellous time. I arrived back in London on the 12th, met Group Captain MacNamara again and was taken to lunch at the United Services Club. There I had an aperitif with Cyril Newall, Chief of Air Staff. A most excellent lunch. I'll have a squadron of RAAF fighters in about a year's time, I think. I arrived back here and immediately on going up, Saturday 20th April, our squadron figured in knocking down five more of these so-called sky terrors. On Monday I got another one. Thank you so much for all the interesting news, I just live for letters and news from home, 'tis gratifying to know you all remember me.'

The Monday referred to was 22 April. The weather was almost perfect, with a clear sky in the afternoon, and the opportunity was seized for training flights for the newer pilots. On the 22nd, Prosser Hanks was leading Flying Officers Lorimer and Clisby, with Pilot Officer Shaw, over Metz. Fifteen Messerschmitt Bf109s were sighted well above, and the Hurricanes could not reach them although Clisby and Lorimer tried some long-range shooting in an endeavour to get the 109s down to fight. Two did swing within range, but when fired on again turned for Germany, and the Hurricanes were too canny to go after them with the rest of the Messerschmitts waiting above. Eventually the two sides swung away for their own lines, and the squadron records referred to the moral victory gained, and the Germans' apparent respect for the Hurricanes and the RAF pilots. There is no official reference to a victory claim by Les Clisby for this combat, despite his own reference in the letter home. He may have been referring to other engagements on 20 and 21 April, when five Bf109s, a Bf110 and two He111 were claimed destroyed, with three other 109s classed as probably destroyed or 'out of control'.

Bad weather kept the squadron grounded until the end of the month and the time was used to send pilots on leave. On 30 April, Squadron Leader Halahan left 1 Squadron to assume command of 67 Wing Headquarters and Johnny Walker became the squadron commander.

There was little action to report in the first week in May. A Bf109 had been captured and was at Amiens, so Halahan led six aircraft over and the pilots inspected the German fighter. Hilly Brown flew it, and engaged Prosser Hanks in low level combat, after which the aircraft's strengths and weaknesses were better understood. Brown then flew the

Messerschmitt to England and the others returned to Vassincourt. On the 8th, the squadron was roused by what the record book called 'alarms and excursions', and was ready for operations at 04.00, but the weather closed down and by 07.00 they were back at 'Available' status. Throughout the 'Phoney War' or 'Sitzkrieg', there had been days of alarm and alert when it was thought the German attack was imminent; nothing had eventuated. Thursday 9 May also was a quiet day; the last.

'It has come.' The simple sentence in the squadron records for Friday 10 May began the account of 1 Squadron during the blitzkrieg in France. The squadron records were lost in the retreat, and reconstructed in England from the recollections, log books, et cetera, of those who returned.

A Dornier Do215 reconnaissance aircraft was shot down near Longuyon by Walker, Brown, Kilmartin, Richey and Soper, then Hanks and Lewis destroyed a Dornier Do17, and Mould and Billy Drake each shot down a Messerschmitt Bf110, but Mould's aircraft was damaged and he staggered back to base where the Hurricane 'passed out', as the squadron record stated. In later combats, Walker had to make a wheels-up landing east of Verdun and Lorimer bailed out of his burning Hurricane.

Great flights of Luftwaffe aircraft were passing overhead as the German offensive broke over the Western Front. The Luftwaffe commanders in the West, Generals Albert Kesselring and Hugo Sperrle, together commanded Luftflotten 2 and 3, which had available 1,848 Dornier, Heinkel and Junkers twin-engined bombers, 468 Junkers Ju87 'Stuka' dive-bombers, 42 Henschel Hs123 close-support planes, 1,440 Messerschmitt Bf109s and Messerschmitt Bf110s as fighters. This formidable force of modern aircraft flown by well-trained, aggressive and confident crews overwhelmed the French Air

Force and the British squadrons. In addition to the Luftwaffe combat aircraft above, there were 576 coastal, reconnaissance and transport planes carrying out their own tasks.

In comparison, the RAF had 26 squadrons in France, totalling 416 aircraft, including 96 fighters in the six fighter squadrons. The French had 564 fighters, 100 bombers, 143 reconnaissance and 260 observation aircraft. The composition of the French Air Force clearly shows its defensive and Army co-operation role.[2]

Airfields and transportation targets were being attacked heavily by the Germans, and it was assumed that they knew of the squadron's presence at Vassincourt, so the unit moved back to Berry-au-Bac, near Reims. However, only half-an-hour after arrival at Berry-au-Bac, the squadron was bombed. The pilots were standing about watching a formation of Germans passing overhead, when one was noticed to leave the formation and begin what was obviously a bombing run. Paul Richey recognised the small changes of direction and said it was a bombing-run, but Les Clisby loudly refuted this, adding that he would not duck even if the German dropped bombs close by. A few moments later the distinctive whisper-whistle-scream of arriving bombs changed many minds about taking cover and all threw themselves flat as a stick of 14 anti-personnel bombs walked in a line of explosions across the clearing towards them. The last bomb was quite close, but no one was hurt. Unfortunately, three French farm workers were killed by the first of the bomb explosions and their badly injured horses had to be destroyed by Halahan and Clisby.

Halahan, back with the squadron, believed that the bombs had been aimed at the concrete building used as a pilots' room on the previous visit, as it was easy to see and the bombs had been quite close. An operations tent was erected on the edge of the woods, with desks and a telephone line to Wing. Trenches were dug and the more thoughtful noted a nearby dugout, in case of future need. On orders from 67 Wing, the squadron remained on the ground, to be available to escort RAF bombers. The French Air Force was responsible for intercepting the Luftwaffe.

On this day, or the next, Les Clisby was mentioned in some records as having destroyed two Dornier Do17s, at Avaux, and landed at the airfield there, but was hit by AA fire from French guns as he was doing so. The squadron claimed five other victories, for the loss of two Hurricanes and no pilots killed or injured. Ten Dorniers were lost from KG2 and KG76 — two at least from KG2 to 1 Squadron — on this day. KG3 and KG77 lost a total of ten Do17s, most to the French and a Belgian fighter.

The RAF in France flew a total of 208 sorties, engaged in over 80 combats, and claimed 42 Germans destroyed for the loss of four Hurricanes. The wreckage of 16 German aircraft was found before dark, but such counting on following days was not possible. Significantly, in view of Dowding's misgivings about his fighter force being bled away into the battles over the Continent, 501 Squadron flew to France for operations during the day. AASF attacks on German forces had resulted in 40% losses and the Bomber Command sorties had lost 9%. In a disastrous Fighter Command attack on Waalhaven, five of six Blenheims were shot down by Messerschmitt Bf110s. RAF losses for the day were 30 bombers and six fighters.[3]

In England, Neville Chamberlain had resigned, and was replaced as Prime Minister by Winston Churchill. As well as France, the Germans had attacked Holland, Belgium and Luxembourg. The German Wehrmacht was about to demonstrate that it had learned the lessons of the First World War. By comparison, the French commander, Gamelin, did not have a radio at his headquarters, having stated that it was not necessary; liaison officers and despatch riders would bring information and take orders to subordinate formations.

The Luftwaffe started 11 May with a well-executed attack on the base of 114 Squadron. At 05.45, the squadron Blenheims were nicely lined up when 12 Dornier Do17s came up the Marne and rained bombs across the airfield. Six Blenheims were destroyed, the other six were rendered unserviceable, the offices were damaged and petrol stores set alight. One Do17 circled while film was shot for the German newsreels.

The morning at Berry-au-Bac was uneventful, but the afternoon patrols were in combat. The squadron role was changed to include defence of RAF bases or headquarters — after Reims had been bombed. At squadron level, there was some quiet enjoyment at the plight of the headquarters officers forced to take cover.

At 17,000 feet north-east of Rethel, at 15.15, B Flight engaged a formation of Bf110s, claiming three. Prosser Hanks was leading, with Clisby as No.2, Mould as No.3 and Pilot Officer Stavent as No.4. There were many enemy formations across the sky, some being engaged by other Hurricanes, who broke off as Hanks' formation approached. From 17,000 feet, Hanks led his flight down onto three Messerschmitt Bf110s three miles (5km) north-east of Rethel. He fired at one, hitting it, bringing white smoke from the starboard engine, and Mould attacked another, also hitting the starboard engine. Clisby then attacked the same 110, hitting the port engine. Stavent fired at the third 110, hit it and saw large pieces falling from the fuselage and tailplane. Hanks and Mould returned to the second Messerschmitt, which dived with both engines out of action, and landed a mile west of Chemery. The crew of two, Leutnant Walter Maurer and Unteroffizier Stefan Makera, 3/ZG2, were taken prisoner.[4]

Clisby and Stavent attacked the third 110, which was last

seen diving at 10,000 feet, one engine stopped, the other streaming white smoke, and was found crashed south-east of Rethel. Clisby went for the Messerschmitt leader, fired at it, and sent it down with the starboard engine and wing in flames, but it was not seen to crash.

At 19.15, A Flight engaged an estimated 40 bombers with an escort of 15 Messerschmitts at 7,000 feet over Mezieres. The Hurricanes attacked the 110s as the bombers swung away north, and the squadron record stated that 'a good time resulted.' The record added that it was difficult to get coherent accounts of the combat as the pilots were still pressing imaginary buttons and pulling plugs an hour after landing. However, apparently eight wrecked 110s and a Hurricane were found. 110s were claimed by many pilots, for a total of 15. Luftwaffe records show two 110s from V/LG1, two from ZG2, two from ZG76, and one from ZG26 were lost. One account has it as on this day that Clisby destroyed the two Dorniers and in some accounts of the campaign, Clisby was also credited with three Messerschmitt Bf109s and a Heinkel 111, which landed in a field. Clisby landed alongside, chased the crew and took them prisoner, an exploit which included a flying rugger tackle to bring one of the Germans down. The French later took the Germans into custody. Owing to the loss of records, it is not clear on which day this highly unusual event took place, but take place it did. Possibly the Heinkel was flown by Unteroffizier Hoffman, KG54.

Wing Headquarters had admonished the fighter squadrons for simply scrambling after passing air raids, and insisted that the squadron await orders for specific tasks. This they did, with some disappointment. During the day, more fighter squadrons had been committed to the battles, with 15 Hurricanes lost for claims of 41 Germans destroyed. However, 29 RAF bombers were lost.[5]

Leslie R. Clisby DFC

Squadron Leader Halahan led on 12 May, when 1 Squadron took part in the disastrous attack by the Fairey Battle formations on the Maastricht bridges, over which German columns were pouring. As well as the many Luftwaffe fighters, the slow bombers were engaged by the numerous German anti-aircraft weapons, soon to be known around the world by the German abbreviation 'flak'. An estimated 300 anti-aircraft weapons were in and around the target area; the bombers were massacred. 12 Squadron was tasked to attack a bridge at Veldwezelt, just west of Maastricht. Volunteers were called for, and the entire aircrew strength stepped

Flying Officer Leslie Clisby DFC RAAF. (Mary Eckert)

forward, so it was necessary to draw lots. Only five Battles were ready and they set off in a formation of three and a two. The sole survivor was the leader of the second element.

On 1 Squadron, Halahan picked the pilots for the mission. His own formation of three included his usual wingmen, Hilly Brown and Clisby. Another five would be flown by Soper, Kilmartin, Lorimer, Boot and Lewis. He intended to arrive in the target area three minutes before the bombers were due, at 09.00, and draw off any defending fighters. They flew past many small formations of 109s, who did not press home attacks on the RAF fighters until they reached the bridge. Then, with one minute to go, Halahan led the Hurricanes on a wide circle, hoping to hold the attention of the Messerschmitts and allow the bombers to attack without interference.

None could imagine the sheer intensity of the flak which was about to fire in earnest. The surviving Battle pilot last saw the leading three aircraft, after bombing, enveloped in a cloud of gunfire, and had no idea of the fate of his own wingman. He had seen Hurricanes trying to hold off about 30 Messerschmitts which attacked 20 miles (48km) from the target, but more 109s attacked closer to the bridge. The Battle pilot dived from 6,000 feet to 2,000, released his four 250-pound bombs and somehow escaped. Hits were claimed on two bridges. In other attacks on bridges and road junctions, 10 of 24 Blenheims from Bomber Command, in Britain, were lost. The defences were so intense that no crew reported bombing results, and fighter combat over the target area also was so intense that the RAF Battles were not seen by all the Hurricane pilots, though the explosions of their bombs near the bridges were noted. However, as well as repairing the damaged bridges, the advancing Germans were constructing pontoon bridges, over which their mobile forces were pouring west.

Halahan's eight Hurricanes were fully engaged over the

target area, and he claimed a 109 and an Arado, but had to land wheels-up; Lewis claimed a 109 but was hit by another, and bailed out of his burning Hurricane; Les Clisby shot down the 109 responsible; Kilmartin destroyed a Heinkel He112 and afterwards insisted that it was a He112, not a 109 and, if so, it is the only known instance of He112 in combat. The experienced Hurricane pilots were probabably given a small advantage by the sheer number of Luftwaffe aircraft in the area, which would tend to crowd each other and whose formation leaders would be distracted by the many other formations nearby. After climbing out of his aircraft, Halahan witnessed a tank battle between German and French units, and saw the refugee-clogged roads at first hand.

On return to the squadron, he remarked that the Belgian troops were running faster than the refugees. When Lewis landed by parachute, his identity was in doubt, so he was jailed for some time but later driven back to the squadron.

Les Clisby is noted in the reconstituted squadron records as having destroyed two Henschel Hs126 battlefield reconnaissance machines and a Bf109, and on this day was awarded the DFC. Contemporary accounts give his score by this time as two Dorniers, two Messerschmitt Bf 109s and two Bf110s, two Heinkel He111s, two Henschel Hs126s, and three others not specified, for a total of 11. Michael Shaw's history of the squadron, written with the assistance of squadron survivors, gives a total of 17 victories claimed on this day, six of them to Leslie Clisby: three 109s and three Arados (probably Henschels). Killy Kilmartin apparently watched, and heard over the radio, Clisby close in behind a formation of 'Arados' and shoot down the last three. Halahan, Lewis and Soper force-landed or parachuted and Brown's Hurricane was damaged. Luftwaffe records include the loss of five Hs126 over Belgium, while I/JG27 lost three 109s over Maastricht. Eleven other 109s were lost that day.

The speed and extent of the German breakthrough had

created a shockwave which rolled through the rear areas, and gave rise to what the squadron termed 'most fantastic rumours.' The sudden sustained frenzy of activity resulted in what the squadron described as 'people who for the past six months had worn spurs to keep their feet from slipping off the desk moved about with some alacrity. Wing Headquarters retired in a cloud of dust to the cellar, due to a bomb dropping on the station opposite.' During the day, the six fighter squadrons of the RAF Component flew 110 sorties, lost five Hurricanes and claimed 18 Germans destroyed, while the three Hurricane squadrons in the AASF lost three fighters for claims of 21 enemy destroyed. Total RAF losses were 36 bombers and 13 fighters.[6]

On 13 May, Prosser Hanks and B Flight destroyed a Heinkel He111, which went down into a field already containing a crashed Bf110 from the combats on 11 May. Lorimer was hit again and force-landed at St Loupe-terria. In another combat, Mould, Goodman and Clisby claimed three He111s and two Bf110s. It is on this day that the squadron record states that Clisby landed alongside the He111 and ran after the crew. The record stated he 'chased the startled crew all over the countryside. He wanted their autographs!' Michael Shaw describes combats resulting in only two victories, for the loss of Billy Drake, wounded. Luftwaffe records show losses of a total of ten He111s on this day.

Fighter Command squadrons fed into the battles during the day lost five aircraft while claiming five, while the two divisions of the BAFF claimed 17 enemy. Total RAF losses were five bombers and ten fighters, of which five were Defiants, savaged by Messerschmitts. Dowding's fears now were reality, as his squadrons were taken from Britain and sent to France, with little benefit to the overall situation. The German flood was spreading and after only a few days the French were already reeling.

1 Squadron lost some drinking and recreation time in the

evening, responding to reports of expected German paratroop landings, and to a false report of signals fired for the benefit of low-flying German reconnaissance aircraft.

Noel Monks related how Les Clisby came to visit him on that evening, after five exhausting flights during the day, and enjoyed a hot bath while Monks gave the well-worn dark-blue RAAF uniform a brushing, and attempted to make it look more presentable. He suggested that Clisby purchase another and pretend that it was the original, rather than buy an RAF uniform, to which the reply was that the original would 'see me through.' This remark could imply that Clisby felt he would not survive; the fact is that he had already ordered—and may have received—the RAF uniforms, including shirts, cap, Mess dress and so on, with a large trunk to carry it all, to the value of 42 pounds, from J.G. Plumb & Son, in London. With a dozen or more victories, the promise of a flight command and later a squadron, it is more likely that Les Clisby was looking forward to the future.

On 14 May, the fifth day of the blitzkrieg, RAF bombers again tried vainly to stem the advance. Squadrons from England and France flew against the bridges, pontoons and advancing columns of German units. Of 45 sorties flown against the river crossing sites, 29 were shot down; of 26 against the roads, 15 were lost; of 109 against the bridgehead at Sedan, 45 were lost. The peacetime RAF light bomber force was simply melting away before the flak and Messerschmitts.

1 Squadron's A Flight had flown against Ju87 Stukas escorted by 109s, landed, and gathered back at the squadron accommodation, leaving B Flight at the airfield. Kilmartin, Soper, Brown, Clowes, Stratton and Palmer had all claimed 109s or Ju87s, for a total of nine. Luftwaffe aircraft were attacking targets in the area, but B Flight was not ordered off.

A formation of some 15 Bf110s flew over the airfield,

about 15,000 feet up, and an engine fitter had run to Prosser Hanks and urged him to take off and 'get into' the Huns. Hanks replied that the 110s were too high and too fast to be caught, but the fitter called on Hanks' pride in 1 Squadron, so reluctantly Hanks looked at Mould, Lorimer and Les Clisby; all nodded.

They went to their aircraft and while the crews were strapping them in, a flight of Heinkels flew over, bombing and machinegunning, but the 1 Squadron airmen carried on and the five Hurricanes of B Flight took off through the attack. They set off after the formation of 110s, but the sky was full of Luftwaffe aircraft.

They engaged a formation of Bf110s, estimated as 30 by Mould, and he had a difficult time getting away from some aggressive German pilots. Prosser Hanks claimed two and Lewis, Mould and Boot, who was with them, claimed one each. A shell hit the forward tank on Hanks' fighter, it caught fire, and he spun down but landed safely. He had been drenched in glycol coolant, and returned to the squadron with what the record described as 'eyes like poached eggs'. Les Clisby and Lorimer were shot down in this combat, and were initially posted as 'missing' but later they were believed to be killed. Les was credited with two enemy aircraft in the action.

Les Clisby must have been aware that the Allied situation was worsening, but died not knowing that the Dutch had ceased fighting; that Rotterdam had suffered greatly in a Luftwaffe bombing attack which took place in error; that the Germans were breaking through on the Moselle; in another fortnight, the evacuation from Dunkirk would be well under way. The long years of 'nothing but blood, toil, tears and sweat' had begun.

Michael Shaw states that Clisby shot down the Heinkel and chased the crew on this day and had the final combat with the Bf110 formation on 15 May. For 14 May, he gives a total of 16 victories to the squadron, for the loss of Lorimer's Hurricane, forcelanded. For 15 May, he gives a total of 13 victories, against a loss of Clisby and Lorimer killed, and the aircraft of Hanks, Walker and Richey lost. For the loss of five pilots — two killed, two wounded, one captured — the squadron claimed at least 87 victories in the first ten days of the blitzkrieg. Shaw believes this figure is conservative, but reasonably correct.

In a conversation the night before he was killed, Les Clisby and Bull Halahan, two uncompromising men who had endured intense fighter combats together and rejected exaggeration and anything false, established to their own satisfaction that Clisby's total of victories was probably 19, of which 14 had been destroyed in the previous three days. After Clisby's death, Halahan calculated that his final score may have been over 20.

Luftwaffe losses on 14 and 15 May, which may have been inflicted by 1 Squadron, could have been Bf109s of III/JG2 and I/JG53, Bf110s of V/LG1 and I/ZG52, Ju88 and He111s of I and III/KG55 and Ju87s from I/St.G77. On 14 May, a total of five Bf110s were lost from LG1, ZG52 and ZG76. 54 RAF bombers and 17 fighters had been lost. Werner Molders claimed his eleventh, a Hurricane, late in the afternoon. The French also were suffering greatly as the superior Messerschmitts ravaged their formations. In the evening, Premier Paul Reynaud asked for another ten RAF fighter squadrons, and repeated the request in a telephone call to Prime Minister Churchill next morning.

On 15 May, Air Chief Marshal Dowding produced figures which clearly showed the British War Cabinet that Fighter Command was being siphoned to destruction in France. The decision was made to stop the flow of squadrons, though more aircraft and men were despatched. By the end of the campaign, of the 260 Hurricanes sent to France, only 66 flew back to England. Not all the others had been destroyed in combat, but were damaged and abandoned on airfields before repair.

Post-war investigation by the RAF found that the French had retrieved Les' remains from the burned out Hurricane, and buried him under a marker as 'Glisby'. It was found that for some reason, he had been wearing wrong or incorrectly marked identity discs, and there was uncertainty as to the identity of the remains, but eventually they were accepted by his mother as those of Leslie Clisby and he was re-buried in one of the Commonwealth War Graves cemeteries at Chuloy near Nancy.

The 'ifs' of warfare often give rise to heated debate, but it may be worth pondering Leslie Clisby's possible career if he had survived that last combat, and the Battle of France. He obviously had become a deadly shot and combat pilot and, had he survived, Clisby may well have doubled his score by the end of the Battle of Britain. Squadron Leader Halahan, however, had most of his 'family' from 1 Squadron with him at OTU, and few of them scored during the battle. Clisby, later, would have had a squadron, then a wing, and may well have ended the war ranked among the leading Allied fighter aces, and quite probably the leading Australian ace. With his engineering background and pre-war graduation from Point Cook, his post-war career almost certainly would have taken him to air rank. The RAAF lost many men of great potential throughout 1939-45, but especially so on 14 May 1940.

Leslie Clisby's score of enemy destroyed, most in the five days 10-14 May, reflects the circumstances which allowed many Luftwaffe aces to accumulate the claimed victories which have caused some controversy among Allied pilots and historians. During this time, Clisby was flying a machine of which he was master; it had formidable firepower; he had, importantly, opportunities to score; he was over friendly territory and, like so many others in the Hurricane squadrons, he could have made his way back to fight again if shot down, but survived. An extreme example is Major Georg-Peter Eder, who arrived in JG51 in December 1940 and flew to the end of the war, but was shot down 17 times, wounded 12 times, and claimed 78 victories, including 36 four-engined US bombers. In all theatres, Allied pilots were presented with spasms of action which allowed some of them to score steadily and heavily, and at other times there was little or no opportunity. As an example, Clisby's brother-in-law, Reg Eckert, flew Spitfire IX and XVI with 127 Squadron RAF from early June 1944 to almost the last week of the war, on many fighter sweeps, bomber escorts and ground attack missions, but in 10 months of operations he did not engage enemy aircraft. On the other hand, Luftwaffe fighter pilots were continuously presented with opportunities to engage in combat and, if they survived the crucial early period, went on to become high-scoring aces or 'Experten'.

Squadron Leader Halahan wrote to Mrs Clisby:

> I have at last received your letter dated June 8th. Our departure from France was rather a hurried affair and the mails have been held up for a long period.
>
> I have been meaning to write to you for some time, but unfortunately most of the squadron's documents were lost in France, together with your address.
>
> Your son Leslie was not only one of the finest officers I've ever met, but my greatest friend in the squadron. We were billeted together from the time we landed in France early in September and we flew together a great deal. As I think you know, Leslie was involved in quite a number of combats before the German Push started on May 10th, and he had been

concerned in the destruction of six enemy aircraft before that date.

From May 10th onwards he was in the thick of the fray, and before he was reported missing he was personally responsible for the destruction of a further eight enemy aircraft.

On May 12th, he was a member of a patrol of eight Hurricanes which I took up to Maastricht to protect our own bombers, which were attacking some vital river bridges. On this ocasion we were in combat for some time with very large numbers of enemy fighters in the target area. One of our pilots was shot down and had to take to his parachute. Leslie noticed he was being shot at during his descent, so he attacked and destroyed the enemy fighter concerned. During this combat, Leslie shot down two enemy fighters and was one of four of my patrol left when the sky had been cleared for our bombers.

During his return from this trip, when his aircraft was damaged and was short of ammunition, he attacked and destroyed two enemy dive-bombers.

On another occasion when he had become separated from his flight during a combat with enemy fighters, he sighted and attacked a large formation of enemy bombers. He shot one down and landed beside it as he saw the crew escaping into a wood. He followed them and disarmed them at the point of his revolver, later handing them over to the French police. His aircraft had become too badly damaged to fly home, so he turned up at our temporary mess some hours later on a French refugee cart, in the best of spirits, having shared a large bottle of red wine with the peasant and his family.

During the four days following May 10th, Leslie was involved in at least six combats, and in each one he displayed that brand of courage and determination which I have grown to associate with Australians serving with the Royal Air Force. For his courage and endurance I recommended him for a decoration and I am very glad to see he was awarded the DFC. I expect by now you have received this decoration.

Of Leslie's last combat I can tell you very little. He was a member of a flight of five Hurricanes which attacked a large

Leslie R. Clisby DFC

formation of enemy bombers escorted by fighters. Three of this patrol were reported missing, but the flight commander returned to us later. I can only surmise that Leslie was taken by surprise from behind while he was attacking a bomber. During the trying days which followed the German advance, Leslie's endurance was an example to us all, and at times when some nerves were a bit frayed, his sense of humour was a Godsend. It was perhaps his ability to see the funny side of everything as much as anything that endeared him to the whole squadron.

I have arranged for two officers who knew Leslie well to go through his kit and I will send you any letters that we have which arrived from Australia after he was missing. In addition I will send you anything amongst his personal belongings which I think you may value, and if there is anything you would like done in the connection with the disposal of his kit, I will see that it is arranged.

As I say, Leslie was my greatest friend in the squadron, a unit which was always more a family than a squadron. I am lucky enough to have most of my old pilots still with me as instructors at this Air Flight School, and they have asked me to take this opportunity of expressing their very deep sympathy with you in your great loss. For myself, I can only say that I feel his loss very keenly and that I realise how you must feel.

I only hope that as instructors here we will be able to teach the young pilots who go through our hands to live up to the fine tradition of the Royal Air Force which Leslie has helped to make.

Squadron Leader Halahan kept as a memento the silver cufflinks engraved with Clisby's initials, and wore them to important RAF dinners, so that something of Leslie Clisby would be present. These cufflinks had been given to Les as a farewell present by his sister Mary, on Les' departure for Britain. Over the years, Halahan corresponded with the Clisbys, and on a visit to England after Halahan's death, Mary was pleased to have the cufflinks returned to her by Mrs Halahan—43 years after Leslie was posted missing.

The memory of Flying Officer Leslie Clisby DFC is continued on the pipe organ of the Walkerville Methodist Church, Adelaide. At St Andrew's Church, Walkerville, his name heads the list inscribed on that church's pipe organ of 15 men who were killed during the Second World War. His name appears on the War Memorial in Adelaide, there is a Clisby Way in Matraville, Sydney, close to streets named for the other great fighter pilots Bluey Truscott RAAF, Paddy Finucane RAF and Cobber Kain RNZAF. His name is in the Memorial Book at St Andrew's Cathedral, Sydney, in one of the Memorial Books in the RAF church in London, St Clement Danes and on the Roll of Honour at the Australian War Memorial in Canberra.

Les Clisby and others who were killed early in the war as members of the RAF were no longer members of the RAAF, and so were not eligible to be included on the Roll of Honour at the Australian War Memorial. This anomaly was not rectified for some time after the end of the war.

As 1 Squadron lost most of its records in France, there is little detail of its part in that desperate campaign, and as the Air Ministry — or its civil servants — decided to destroy the many log books accumulated during the war years, without contacting next-of-kin or families, it is not possible to include information from Leslie Clisby's log book. Individual aircraft flown by Clisby have been identified in previous pages, and it is known that L1694 NA*F was flown by him in France.

Of the pilots who flew with Leslie Clisby pre-war and in France, casualties were severe. 'Hilly' Brown became commander of 1 Squadron, and accumulated 18 victories but was killed on 12 November 1941 when leading a wing attack from Malta on targets in Sicily; Billy Drake went on to accumulate a score of 24.5 victories in the air, plus many more as probables and a further 20-odd destroyed on the ground;

Squadron Leader Patrick 'Bull' Halahan survived the war; Peter Prosser Hanks survived the war, with at least 14.5 victories; Caesar Hull eventually became commander of 43 Squadron on 1 September 1940 and was killed when, with no ammunition left, he went to the aid of Dick Reynell, the Australian test pilot with Hawker who was gaining operational experience, and Reynell also was killed in the combat; Lorimer was killed in the same combat as Clisby; Peter 'Boy' Mould, with nine victories, was shot down and taken prisoner when flying from Malta; Pennington-Leigh, whose father owned the boat on which Les Clisby so enjoyed his leaves, was killed in action in Burma, as a squadron commander; Paul Richey survived the war with at least ten victories; Peter 'Johnny' Walker also survived the war, with a score of 8.5 victories.

Bader's bus company

Fighter Sweep Spitfire

Sqn Ldr Tony Gaze DFC**

SIX ACES

Tony Gaze DFC

Tony Gaze flew Spitfires from early 1941 to almost the last weeks of the war in Europe and then flew the Meteor jet fighter until war's end. As such, his flying career may be unique for an Australian fighter pilot; he was almost certainly the first Australian operational jet fighter pilot. Tony was born on 3 February, 1920. Some 20 years later, he was in England, at Cambridge, when the Second World War broke out. The recruiting teams came around, and the Australians present gathered for a meeting to decide what to do. Some wanted to return to Australia and join in their own nation's forces whilst others saw little point in this and decided to enlist at once, albeit in the British armed forces.

However, in the early days there was no organisation able to absorb the large numbers of volunteers for the RAF and, like many others, Tony Gaze was told he would have to wait. Then his brother, Scott, in Australia, sent word that he was coming and to wait so they could join together. This they did, with a little juggling of age dates for Scott. Tony believes that the probable clinching factor in their acceptance was his statement that his father had been a fighter pilot in the First World War and he wanted to be one in the second.

The RAF still did not have training facilities able to cope, so the two Gaze boys, with rank of Aircraftman 2nd Class (AC2), swept, cleaned and laboured as ordered. It was then discovered that both had been in the school cadets and had some knowledge of weapons, so both were sent to the north of England as airfield defence gunners in Lewis gun teams. They also flew training flights as gunners—in the days when there was no flying brevet, rank or pay for this work—with 10 Squadron RAF from Dishforth when that squadron was engaged on leaflet-dropping missions over Germany.

Finally, in May 1940, they began flying training. They were based at Cambridge and, as a former undergraduate,

Tony was able to dine in the Hall, but the rest of the time was spent with the other cadets. Later they moved to Sigrist's pre-war training school, which was quite comfortably established, and operated 36 Tiger Moths. Their days were enlivened when fear of an invasion led to fitting bomb racks onto the Tigers, but this was not seriously considered after Tony recalled 'a very frightened Chief Flying Instructor came down after the first test with the racks.' Spin recovery was extremely difficult, as the racks altered the aircraft's centre of gravity. However, Scott and Tony did some dive-bombing practice, before moving on to the Miles Master at Sealand. Like many other young men in flying machines, Tony indulged in unauthorised low flying, resulting in his log book being endorsed in red ink by the AOC 21 Group for the misdemeanor on 30 October 1940. During their training, the cadets could hear the Luftwaffe droning overhead on the way to Liverpool or Chester, and perhaps see the distant activity.

Well aware of the potentially disastrous effect of bombing on urban areas, Hitler had categorically refused to allow such attacks on British cities, and forbidden bombing of London. But on the night of 24/25 August, Luftwaffe bombers accidentally hit London. Churchill ordered a retaliatory strike on Berlin and the fate of hundreds of thousands of people was sealed—urban areas became targets. Hitler and Goering were infuriated, demanded a full report from the Luftwaffe, for the crews to be identified and punished and for each responsble aircraft captain to be posted to an infantry regiment.

The RAF attack on Berlin on the night of 25/26 August was a failure, but the fact that enemy bombers had arrived in the area of the capital of the Reich was a severe blow to civilian morale and to the statements by Hitler and Goering that such attacks would not occur. Hitler was forced to retaliate in turn, with a night attack, and so bombing cities became an accepted part of military operations in the Second World War. On 7 September 1940 the first great attack on London left 436 people dead and 1600 injured; the Luftwaffe returned for 57 consecutive nights. Coventry was a manufacturing centre with nearly 60 legitimate military targets, and suffered heavily on the night of 14/15 November, with 554 people killed and 865 injured. The centre of the city was destroyed, and 'to Coventrate' was used to describe similar destruction. Night defence against bombers was almost non-existent, and the Luftwaffe struck across the length and breadth of Britain: Birmingham, Bristol, Cardiff, Liverpool, Manchester, Plymouth, Sheffield, Southampton, Swansea and many other locations suffered heavily. By mid-February 1941 over 700,000 homes had been destroyed or damaged.

On 8 January 1941, Tony graduated from No. 5 Service Flying Training School as an 'Above Average' pilot, with just under 122 hours in his log book. February and the first week of March were spent at 57 Operational Training Unit, Hawarden, where he amassed a total of 15 hours 15 minutes on the Spitfire. As Hawarden was too boggy, they would go by bus to Speke, Liverpool's airport, which was not quite as soggy. There was a sweep running as to the greater number of aircraft wrecked: Spitfires stood on their noses by trainees unable to cope with the mud, or American Curtiss P40 Tomahawks from the assembly plant ditching in the Mersey with failed Allison engines.

Tuition at 57 OTU was 'very haphazard, no real instruction at all,' Tony recalled, 'we were just sent off to enjoy ourselves flying Spits around the sky.' The first time he looped a Spitfire, he blacked out; everything else he had flown had required definite pressure backwards on the stick but the nimble fighter reacted at once.

At the end of his time at 57 OTU, and for the first and only occasion, Tony was given what he requested in the RAF. As his father had flown from Tangmere in the First World War, he asked for a posting there and got it, going as a new Pilot Officer to 610 Squadron at Westhampnett. He now had a total of 140 hours 25 minutes in his log book.

Spitfire MkII DW·G, 610 Squadron RAF, in which Tony Gaze claimed 3.5 enemy destroyed, 2 probably destroyed and 1 damaged. (Tony Gaze)

As a member of A Flight 610 Squadron, Tony flew many training and orientation flights, with two uneventful operational patrols up to the end of March, and these continued to the middle of April. His brother, Scott, was a member of B Flight, but was killed on 23 March 1941. A Ju88 was being chased in bad weather, and Scott was found in his crashed Spitfire, gunbutton to 'Fire', but it was not known if he had engaged in combat.

Tony found that operational squadrons were quite different from training units. 'The first time you flew with your flight commander, when it was time to go home he'd just flick onto his back and go straight down, the sort of thing you'd never think of in training, to see if you'd stay with him, with everything right off the clock.'

610 Squadron had been combined with 145 and 616 to form a wing led by the famous Douglas Bader

> who completely altered our idea of a combat formation. After giving us a shock when we first met him, because he was such an outgoing character and we thought him a bit of a bull artist, we flew with him and we thought he was the best leader we'd ever known. I still do. He'd hold everyone together and read the way a battle was going. He'd give a running commentary, and if anyone sounded worried he'd buck them up, and always go back and help anyone in trouble.

The RAF determined to take the war into France and Holland. A few bombers were to be sent to bomb selected targets, in the hope that the Luftwaffe would come up to fight and engage the large escort of Spitfires and Hurricanes. Bader had attended a briefing at Headquarters at Uxbridge, where the basic idea was explained to him and the other assembled leaders by Air Vice Marshal Leigh Mallory. This type of operation was called a Circus, and the first was launched on 17 April 1941. Bader led the wing as part of the Spitfire escort to a dozen Blenheim light bombers which attacked Cherbourg. There was no Luftwaffe reaction and Bader was disappointed.

Tony Gaze also was disappointed, for he did not fly on the operation, but did an R/T test in Spitfire DW-A, saw a bulky twin-engined aircraft and identified it as a Beaufighter, later finding to his annoyance that it was a Junkers Ju88, and followed the notation in his log book with a short string of !!**##.

More training and operational patrols followed, including an escort to Blenheims bombing Le Havre on the 21st and then, on 24 April, Tony flew his first Rhubarb—a flight of two fighters into France attacking anything they encountered. No enemy aircraft were seen, but barges and guncrews were strafed. More bomber escorts followed, with the usual patrols, to the end of April. May also was quiet, with patrols and bomber escorts. On three sorties enemy aircraft were chased, but there was no combat. This was disappointing to the aggressive fighter pilots and their leaders, like Bader and Sailor Malan, because all available information was that northern France was well covered with Luftwaffe bases.

The night attacks on British cities continued and while exact details of civilian deaths, destruction and suffering were not released by the authorities, it was obvious great damage was being done. To the end of March 1941 nearly 29,000 British civilians had been killed and 40,000 injured by the bombing or 'The Blitz' as it was called. In April, another 6,000 civilians were killed. On 19 April, the Luftwaffe flew 783 sorties over London which dropped, for the first time in air warfare, 1,000 tonnes of bombs on a target. Twelve hundred Londoners were killed and 1,000 injured. On 8 May, 450 people died in a raid on Liverpool; on 10 May 500 bombers attacked London, creating over 2,000 fires. RAF bomber attacks were ineffective, but British politicans, the public and RAF bomber leaders regarded the Blitz as something for which Germany would pay in kind. The destruction visited on British cities would be only a fraction of that delivered to Germany, though RAF retribution was some years away.

For the RAF dayfighter squadrons crossing into France, early June 1941 saw the Luftwaffe avoiding contact. At squadron level, no one knew the Germans were preparing to attack Russia and that they left only a small force in France. But in the last half of the month the Luftwaffe began to react more strongly. On the 17th, during a combat in a Circus to Boulogne-Dunkirk, six Messerschmitt 109s were reported on

Tony Gaze's tail, but he never did see them. He did see one in the sea, with rescue boats busy, and Flight Lieutenant Les Knight claimed one 109 destroyed. Next day there was 'bags of flak' and he saw a Spitfire in flames, as well as six 109s above. Sergeant Merriman claimed one destroyed. The docks at Le Havre were hit on the 19th but there was no Luftwaffe interception. The Spitfires were flying two Circus operations a day, and Tony's log book reflects the intensity of the period, with notations such as 'bags of flak', 'one 109 seen', 'saw rescue launch burning', 'bags of 109s'. St Omer, base for the Messerschmitts of JG26, was bombed on 21 June. Flying No. 3 to Les Knight, Tony engaged several 109s, chased one and fired his cannon from 500 yards (metres) in to point blank range, claimed it as probably damaged, and then fired at two more head-on, claiming them as 'very frightened'. However, two other 109s attacked, putting 19 holes in his wings before he broke away, landing at Lympne with an oily windscreen and no air-speed indicator. After another combat, when 109s destroyed were claimed by Squadron Leader Holden, Pilot Officer Horner and Sergeant Rains, the only one of the section not to claim was Tony, who added in his log book, 'what the hell was I doing??!'. On 25 June, he wrote 'D.B. got one' when Bader destroyed a Messerschmitt. Bader, recalled Gaze, 'had this great ability to sound as if nothing was wrong. Even when we were in real trouble, outnumbered, he'd always say something like, "Come on fellas, we'll get some more of these!" It rubbed off.'

Another aspect of the war and of young men in danger almost daily, was the warrior's sense of humour. 'The narrower the escape, the more hilarious everyone thought it was. A fellow would come in, shot to ribbons, scared out of his wits, and everyone thought it the greatest possible joke.'

On 26 June, a Circus went to Lille and in the 610 Squadron formation Gaze was No. 3 to Squadron Leader Ken Holden, the squadron commander. A combat began, and Tony 'picked one 109 off the CO's tail; destroyed' then attacked 12 more Messerschmitts, firing at the one on the outside of the formation, but the others went for him, he dived to sea level to escape, saw a 109 crash into the water and the pilot in the sea. He claimed one destroyed and one probable, adding to the log book entry 'AT LAST'. No German pilots were recorded as killed or wounded on this day, and it is possible both pilots baled out safely. Next day was a 'great shambles', with accurate flak from Boulogne, many 109s who did not attack, but a Spitfire which did fire at him. The summer month ended with more escorts to Comines and Lens, but no other combats, although 109s were seen.

July began with another mission to Bethune — no Messerschmitts but Boulogne's accurate flak. On the 2nd, during a Circus to Lille, Bader said he was going to attack. Gaze followed Squadron Leader Holden in the turn, then saw a 109 300 yards (metres) away slightly behind and below, turned down on it and fired a one-second deflection burst. There was a puff of whitish smoke from the port side of the cowling, and the 109 half-rolled and went down. Gaze followed, firing two one-second bursts at 200 yards (metres), and saw cannon strikes on the 109's fuselage. The German then dived past the vertical and at 5,000 feet Tony broke away and climbed. During the combats, Bader himself claimed one destroyed, one damaged and one frightened after watching the alacrity with which the third 109 whipped over and dived vertically away when fired at. On the 3rd it was another 'shambles of Spitfires and Hurricanes' and a few 109s were seen in the morning Circus, repeated in the afternoon, and with similar operations on the 4th and 5th.

On 6 July, six Stirlings bombed Lille. Again flying as No.3 to Holden, Tony was impressed by the size and bombload of the new four-engined bombers, which dropped 36 tons of bombs onto the target, flinging debris up to 1,000 feet. There were many 109s attacking, and Holden turned after two 109Es, Gaze followed faithfully, Holden attacked the second 109, which began to smoke and broke down. Gaze fired at 250 yards (metres), Holden turned away but Gaze went down after the 109 and fired again at 200 yards (metres) and watched the badly smoking 109 diving vertically until he lost sight of it in the summer smoke haze. Coming back from a Circus to Albert the next day, as No. 2 to Holden, they attacked some German E-boats, saw an aircraft in the sea and transmitted on the radio so a fix could be made of the location to assist rescuers. In the afternoon they escorted three Stirlings to Crocques. Lens was attacked on the 8th, and one Stirling was blown up, then a Spitfire was seen spinning, trailing glycol. Another Spitfire hit the sea and recovered, no doubt with a shaken pilot. Haze and sun made the flying difficult, but no 109s were seen.

By now, the Luftwaffe must have been practised at predicting the target of the approaching bombers and when Lens was attacked on the 10th, Messerschmitts were there in force. 'Hell of a scrap with 109s' Tony noted in his log book, and shot one off Holden's tail. He destroyed two 109s during the sortie; one blew up and the other dived into the ground. He also noted 'very frightened' and landed at Shoreham with four gallons (18l) of petrol in his tanks. All the flights over Occupied Europe were made with one eye on the fuel gauge. The Spitfire was not designed as a long distance fighter, but as a defender of the United Kingdom, for which an hour's fuel was reasonable. The 109 which exploded may well have been flown by Unteroffizier Erich Hammon, of 2nd Staffel I/JG26, who was reported missing, and no trace of the 109 was found by the Luftwaffe. There is no record of other pilot losses, so the second may have baled out.

Attacks on Lille and Hazebrouck followed, then on 17 July

Boulogne was the target again. At 17.40, while flying up the French coast at 20,000 feet, a formation of 109s was seen on the left. Bader turned onto them, and Gaze followed Ken Holden, saw more enemy above the formation being attacked by Bader, and called a warning. He turned onto one of the higher 109s and nearly collided with the German wingman, who dived away with Gaze after him. He fired three one-

Tony Gaze, 610 Squadron RAF, July 1941. (Tony Gaze)

second bursts, seeing a hit on the fuselage from the last burst. By this time the ground was leaping up, less than 2,000 feet away, and Gaze pulled out violently, blacking-out in the process. The 109 was still in a vertical dive 150 yards (metres) in front, and Tony believed he must have gone into the ground but could only claim a probable Messerschmitt 109F. Unteroffizier Gerhard Oemler, 9th Staffel III/JG26, failed to return from this combat, and no trace was found of him or the aircraft. Oemler had claimed two victories. On return to base, Gaze was informed of the award of the DFC, and then 'forced on leave. Didn't enjoy it.'

He came back on 25 July, and flew to the end of the month without further contact. Bader was still leading the Wing and was, in Tony's opinion

> an absolute devil. We used to come home on the deck. We were always trying to get up enough nerve to fly through the holes in the piers, where they'd cut a hole (as an anti-invasion measure) and always chicken out at the end, thinking perhaps (the Spitfire wingspan) would not fit. As a Pilot Officer, I was flying close formation, trying to show how good I was, and he kept making a dart at me, flick his wing at me, to see if I'd duck. (He'd) try you out all the time.

The first week in August passed with patrols and offensive fighter sweeps over France, but no combats, and then on the 9th a Circus attacked Gosnay. 610 Squadron provided high cover for the Tangmere Wing, remaining high while Bader led the others into the 'bags of 109s' below. Spitfires and Messerschmitts tangled, some disappearing into the cloud cover at 15,000 feet, the combats faded and the squadrons returned to England.

It had been Bader's habit, begun as a morale-raising move, to slide back his canopy, slip his helmet off and enjoy a pipe on the return flights. During this time he was off the air, and no one thought it unusual this day not to hear him. He had lost the rest of his section when he had to break violently to avoid colliding with a German then, lower down and alone, he had attacked six more 109s, hit two and tried to pass between two more, but is believed to have collided with one which cut off the tail of the Spitfire. Bader barely escaped in time, came down by parachute was captured minus one of his 'tin legs' and began his career as prisoner-of-war and making life as unpleasant as possible for his captors. Recently, a German pilot claimed he shot down Bader, but the matter has not been resolved.

For the Tangmere Wing, a great blow had fallen. Bader had seemed indestructible. 'We were like lost kids after he went,' said Gaze. 'We were so used to him that the chap who followed him had a hard time.'

However, the war had to go on and the traditions of the RAF, begun with the first Royal Flying Corps losses in France in 1914, did not allow feelings of gloom and worry to affect the operational aspects of a unit. The sweeps and Circuses continued without let-up. On 14 August, leading Yellow Section, Gaze became engaged in 'quite a long dog-fight' with a bunch of 109s, while mopping up after a Blenheim raid. Part of his hood blew off, and at one time some 109s came along and formated on him — aircraft recognition was a problem for both sides.

In the next few days there were operations to Le Havre, St Omer, back to Le Havre, Le Touquet, Lille, Gosnay and Wolques. The Germans had offered safe passage to an aircraft carrying a replacement leg for Bader, but this would have been an obvious gift to Goebbels' propaganda machine, and the leg was dropped during a normal raid, as the aircraft passed over St Omer on the Gosnay mission. Tony saw a Spitfire shot down by a 109 during all this, and privately thought it was going a little too far when all that was necessary was to deliver the leg.

Then on 21 August came what he described as 'the squadron's worst day', during the afternoon sweep to Cholques. The 109s had not appeared on the morning Circus, but engaged the second visit. Crowley-Milling, Black, Wright and Rains were missing. Only Ken Holden and Gaze made it back from their formation, and Tony had two gallons (9l) of petrol left. 'Very frightened' he noted in his log book. On the 26th, another Circus bombed St Omer and Tony ruefully noted 'missed some sitters, 109Fs with yellow (noses) right to cockpit. Did not squirt.' Next day was a repeat performance, and two yellow-nosed 109s did a head-on attack, but again there was no firing and no claims.

On the 29th, 610 moved to Leconfield in 12 Group, away from the daily sweeps over France. During his time with 610 Squadron, Tony had flown 49 fighter sweeps or Circuses, seven Rhubarbs or Sphere sorties, claimed 3.5 destroyed, 2 probables and 1 damaged. He had a total of 334.5 hours.

> We were a bit tired by then, and sent up north for a rest doing convoy patrols. These were frustrating, because you knew you were needed, the Ju88s were on the horizon waiting for you to go, yet were so low you couldn't see them. If you were doing the last patrol you'd wait until night, then set off for home. Before you'd gone five minutes the guns would start and the 88s would be in. There was nothing you could do about it.

After this squadron flying, Gaze was posted on 18 November to 57 OTU, from which he had gone to 610 in March, in his own words, 'made an official "dead beat"'. He had flown 483 hours, with just under 218 as operational hours.

At the OTU he was a 'basic instructor', then moved to gunnery and finally had a flight of his own. There were many RAAF pilots coming through, and he did his best to pass on a better standard of training than he had received. Almost all the people at training units were not instructors but were merely having a six month break from operational flying—

an unfortunate and ill-founded policy. Instructing is an art which requires the full attention and understanding of a person suited to the task. Operationally tired pilots rarely were able to meet the requirements. In addition, being sent to instruct was often regarded as less than a compliment.

During this time, two pilots who later became famous aces were allocated to Gaze: Don Gentile, an American, and George Beurling, a Canadian. Between them, they destroyed about 60 enemy aircraft; both were killed in post-war crashes.

Gaze regarded his time at OTU as wasted—despite flying 305 hours 40 minutes whilst at Hawarden—and was overjoyed to find himself posted on 1 June 1942 to 616 Squadron at Kingscliffe, with Johnny Johnson as his fellow flight commander. Johnson had also flown with the Bader Wing. 616 was equipped with the Spitfire VI, a special high altitude version. Powered by a Merlin 47 fitted with a four-bladed Rotol airscrew, armed with two 20mm cannon and four .303-inch machineguns, the Spitfire VI was also fitted with a pressure cabin which provided the pilot with the conditions of 28,000 feet at 40,000 feet. However, the canopy did not slide, and was screwed into place after the pilot entered—a feature which was not too popular with the men who had to fly them. One advantage of the cockpit construction was that the pilot was relatively well-protected from the initial effects of fire, being in a small cocoon while they could undo their straps and prepare to bale out, unlike the normal Spitfire cockpit, where the pilot's feet were on the fuel tank in the nose. One hundred of this model Spitfire were built and issued to 616 and 124 Squadrons. But no sooner had the RAF demonstrated that it could counter the high altitude threat than the Luftwaffe reverted to low level attacks, and the pilots found themselves sweating in the hot-houses created by the special cockpits as they operated at lower altitudes.

June 1942 passed with a variety of training flights interspersed with operational sorties which did not meet any enemy. In July, 616 Squadron moved to West Malling for what was intended to be the original date of the raid on Dieppe, then to Kenley, and sweeps were flown across the Channel, to the places familiar to Tony from the previous year — St Omer, Dieppe, Le Touquet and Abbeville.

On 13 July, Tony led the squadron on a sweep to Abbeville, fired at a FW190, saw the flash as high-explosive shells burst on the wing and cockpit, then the 190 spun, and was still spinning below 9,000 feet when last seen by three other pilots. It was claimed as damaged, but later changed to a probable after assessment of the cameragun. Oberfeldwebel Helmut Ufar, who had claimed three victories, was found in the wreckage of his FW190 from 4th Staffel of II/JG26, and was reported as failing to bale out of his spinning aircraft. He may well be Gaze's victim, as there was only one other claim on this date.

On 18 July, Gaze was scrambled after two FW190s and chased them to the French coast but was unable to make contact. On the way back to England he saw two other 190s going back across the Channel so he swung to attack. He flew under the leader, pulling up to fire into its belly, and saw the undercarriage flop down, then turned onto the other, damaging it. The first one was in flames, and the pilot baled out some 10 miles (16km) off the coast at Le Touquet — his parachute had four holes in it. These may have been aircraft of JG2, but there is no record of this combat in surviving German documents.

More sweeps were flown and on 30 July the squadron flew from Great Sampford to the Boulogne area. Another squadron was leading and the commander put them all in a position where they were bounced by at least 80 FW190s and Bf109s. Tony described the RAF formation leader as a 'wet prick. No idea at all.' A 109F shot down his No.2, Sergeant Cooper;

Tony turned onto the German, who attacked head-on, they both fired at pointblank range, and Tony distinctly saw the flash from the 20mm cannon in the Messerschmitt spinner, then saw what seemed to be the cannon shell itself appear to fall out of the muzzle and drop away. He assumed the German had a misfire and then they were past each other and turning tightly. The Spitfire began to win the contest and the German swung away, pulling up his nose. Tony swung in behind the Messerschmitt, which was climbing steeply, firing from 300 yards (metres) dead astern, but the Spitfire cannon recoil slowed him quickly, and the Spitfire spun; so did the Messerschmitt. Tony's camera recorded a good picture of the 109 but no other results were known and he later claimed the 109 as 'very frightened'. As well as Cooper, Pilot Officer Large was shot down, though claimed a FW190 probably destroyed, and both he and Cooper were picked up by air-sea rescue launches.

In early August there were more test and training flights with only two uneventful operational sorties, until the landing at Dieppe on 19 August. An assault force, largely composed of Canadian units, was to occupy the port of Dieppe, then conduct a withdrawal, and cross back to England. The operation went badly wrong; the Canadians were defeated, the port was not taken and the Germans were able to boast of another victory against British land forces on the Continent. Four thousand three hundred and eighty men were killed, wounded or captured, and every vehicle landed was abandoned. Thirty-four ships were lost. The air fighting was fierce, with the Luftwaffe reacting strongly. It was the greatest single day's fighting of the air war, with 106 RAF aircraft lost against 170 Luftwaffe aircraft claimed.

Tony flew four sorties on the 19th. On the first, FW190s mixed it with the squadron, and he fired at one but the German wingman was there, firing, so Tony broke. On the second, he destroyed a Dornier Do217, probably from KG2, trying to attack the gathering of Allied shipping, but then 30-plus 190s attacked, and the Spitfires were surprised to be able to outclimb them. When the Spits were in a position to attack, the 190s dived away.

Over-claiming, though in genuine belief of victory, marked almost all the days of pitched aerial battles and on this day the RAF claimed 31 Do217s destroyed, 7 probables and 45 damaged, while KG2 acknowledged a loss of 16, with 7 damaged. On the third sortie, there were few enemy aircraft to be seen, but some attention was paid to Dieppe, 'in flames, thoroughly plastered'. The final flight was covering the shipping evacuating the Canadian assault forces; a Do217 bombed, but missed, escaped, and the squadron returned.

Because of the general disaster, it was believed by many that the Germans had been warned of the raid. A thorough study of this and other myths of the Second World War by Nigel West, in *Unreliable Witness*, shows that there was no warning to the Germans, at theatre command level or lower. The rest of August was devoted to sweeps over northern France and Holland, to Ostend, Rotterdam, Abbeville, Gravelines, Albert and Lille.

Elsewhere in the war, the Japanese had been halted by Australian ground forces in New Guinea and been defeated in the naval battles of the Coral Sea and Midway. In Africa, General Montgomery had taken command of the Commonwealth forces opposing Rommel. The Russians were retreating before the German 1942 summer offensive; General Paulus' 6th Army was ordered to attack Stalingrad.

On 1 September Tony took command of 64 Squadron, with Spitfire IXs at Hornchurch. The first operation was on the 5th, escorting 36 B17 Flying Fortresses to bomb the marshalling yards at Rouen. 109s appeared on the way home, and the

Spitfires circled with them but there was no combat. Next day, escorting 16 Fortresses to St Omer, he led the squadron onto about a dozen 190s, scoring hits on two, seeing cannon strikes on one and white smoke pouring from the other. As as the early Mark IX Spitfires had an oil cooler in the wing instead of a camera, there was no independent record of the successes, although Tony believes that the enemy hit during this period were destroyed. However, the first two B17s had been lost during the mission—one each from the 92nd and 97th Bomb Groups—to attacks by FW190s. American gunners claimed four Focke Wulfs destroyed, with others as probables and damaged.

On 7 September, 36 Fortresses bombed Rotterdam, but took the scenic route, going past Amsterdam and swinging out over the Zuider Zee, back to the target. Some bombed, but others simply jettisoned the loads. The Spitfires landed with 'damn all petrol left.' That afternoon, Tony flew as co-pilot for the short flight to Polebrook, the B17 base, to attend a party. His pilot was Major Tibbetts, later to play a leading role in the atomic bomb missions over Japan. His B29, *Enola Gay*, dropped the atomic bomb on Hiroshima.

Bad weather curtailed operations for some days, until a mixed sweep with newly operational US P38 Lightnings was mounted over Le Touquet-Abbeville-Dieppe, but only one FW190 was seen, there was no flak and no combat.

Then on 26 September, nature took a hand. Seventy USAAF aircraft took off, intending to attack Morlaix. Three squadrons of Spitfire IXs, from 64, 401, and 133 (the American volunteer 'Eagle' Squadron), were to provide the escort. The nominated wing leader, Brian Kingcome, damaged his tail-wheel on landing at the Bolthead strip, missing the briefing, and Tony assumed command. McColpin, commander of 133, was to attend a conference, so that squadron was led by Flight Lieutenant Gordon Brettell RAF, who was known to be somewhat impetuous.

The weather was bad and, unnoticed, a 140-knot tailwind carried the formations well south of their intended track. The bad weather caused most of the bombers to turn back, as accurate bombing would have been impossible. When the Spitfire escort arrived at the intended rendezvous, there were no B17s in sight. They circled, unable to see the ground through cloud then, away to the south, the bombers appeared, struggling back against the wind. The formation of B17s had split in two, 18 turning back but six going on, over the Bay of Biscay, to Bordeaux and almost to Spain. The Spitfires joined the B17s, but then had to contend with the wind on the trip home, and cloud below made navigation and position-finding difficult. After flying back for about the same length of time as they had on the way out, a strip of coast was seen below. But it did not look like the coast of England so Tony called control and asked where radar had the formation located. The reply was 160 miles (250km) south of England. He realised they were over the Brest peninsula. Meanwhile the US squadron thought the land was the south coast of England, rolled over and dived. Too late, Tony called out but they had gone, through the cloud and into the gun-sights of the waiting German anti-aircraft defences.

133 Squadron pulled into a tight impressive formation for the benefit of the watching populace, supposedly British in Southampton-Plymouth-Portsmouth, and passed over the roof tops at about 2,500 feet. The gunners opened fire. Above the clouds, the other squadrons heard radio calls from the Americans, 'Stop these bastards shooting at us!', but it was too late. None of the squadron returned. Those who survived the flak, crashed and became prisoners, moving to Stalag Luft III in Silesia.

Above the clouds over the Brest peninsula, the other squadrons went into a gradual descent, hoping to reach England; as it was, some aircraft ran out of fuel during their approach to land. Below the other two squadrons, struggling back alone, was the sole American survivor, Beaty. He had been slightly wounded, but was able to talk to Tony, who told him to climb to 3,000 feet and press Button D on his radio so that a fix on his position could be taken in England. Beaty had gone low, but had to climb to allow the radio and radar to pick him up from England. Tony realised Beaty was not going to be able to get back to England, but advised him to throttle back to 1600 revs and use as little boost as possible, which would give him maximum duration in flight. Gaze told the other squadrons to fly with 1800 revs and two pounds of boost, which would give them maximum range. Immediately after landing, Gaze asked what had been done about picking up Beaty, but 'it didn't seem they'd done anything at all. I asked if something had gone out, a Walrus or something, and they couldn't tell me anything. They'd had about half an hour to do it, as he'd been talking to me almost from the minute he left France.' Both Gaze and Beaty had been sending Mayday calls for at least twenty minutes. Beaty disappeared into the sea. 64 and 401 Squadrons returned, though with difficulty, and with crash-landings.

In Tony's opinion, '10 Group made a complete mess of it. They should have warned us; our progress across the map must have been absolutely astonishing.' The official history states that only 14 bombers took off, with time over target to be 16.45, the fighters never contacted the bombers, but at 16.50 were plotted 30 miles (48km) south of the target and were ordered to return, which was acknowledged by 'the Wing Leader', who then saw the bombers away to the south and turned to investigate. There was no radar plot until 17.16 hours, when the composite formation again appeared on the screen and the fighters began to go down for lack of petrol. The B17s had not turned back until 20 minutes after acknowledging the recall signal. An investigation by Fighter Command and 8th Air Force found that the three causes of the tragedy were complicated orders from 10 Group, incorrect wind information used by the B17 navigators and failure to arrange joint briefing for the fighter and bomber leaders. It seems no independent checking of wind was done by the bombers or, if it was, the information was ignored. The official history implies that 133 Squadron was lost through lack of fuel, and there is no mention of the formation pass over the city below. Tony and Brian Kingcome were driven through the night in the back of a truck, going to HQ Fighter Command to explain. He was greeted by Air Vice Marshal Broadhurst with, 'Ah, Gaze. I see you're an Australian. I don't like Australians.'

Some Eagle Squadron survivors believe blame for the disaster should be placed on the person responsible for the weather information presented to the crews. Tony Gaze thinks that even when the briefing was given and the force was airborne, the speed with which the formations moved across the radar screens and plotting board should have alerted the duty controller, but apparently no alarm was felt even though the Circus was right off the board. In the end, he received the blame.

Next day, to add to personal misfortune, Tony crashed his personal Spitfire, the beautifully polished Mk IX SH-G, when he had to pull up the undercarriage to avoid over-shooting the runway. The airfield was on a hump, 'and once you rolled over the hump you just started rolling down. There were no brakes or flaps. I tried to ground-loop it, but couldn't, so pulled the wheels up before I ended up in the village.'

So September 1942 came to a dismal end.

October operations began on the 2nd with a fighter escort to six B17s of the 97th Bomb Group, bombing the St Omer dispersals, causing a large fire. Five 190s attacked the bombers as they were crossing the coast, but the Spitfires were unable to engage them. No bombers were lost from this or two other missions flown at the same time against the aircraft factory at Meaulte and as a diversion along the coast. A week later the 8th Air Force mounted its largest raid to-date, sending 111 B17 Fortresses and B24 Liberators to the steelworks at Fives-Lille in Belgium. Inexperience caused a poor bombing pattern and damage outside the target area, with civilian casualties. Two B17s went down over France, and two more into the Channel, with the crews picked up by launch. The bombers claimed 56-26-20, which was almost the figure of the total German defending force, and so began the sequence of USAAF over-claiming and British derision at the figures. US claims were later revised to 21-21-15, but German records show only two fighters lost to enemy action, though it is believed that these records are themselves incomplete. 64 Squadron was part of the Target Support formation. Twelve FW190s were scared away from the bombers, and Colin Gray, the New Zealander, and Gaze stayed in mid-Channel until all the bombers had gone by.

On 11 October, six Spitfire squadrons flew to the St Omer area and at 28,000 feet a formation of FW190s were attacked. A melee resulted, and Tony turned onto one 190, fired an eight-second burst from 300 yards (metres) right in to 50, broke away sharply to avoid ramming, watching over a dozen flashes from his exploding 20mm cannon on the wing roots and fuselage of the Focke Wulf, but the Luftwaffe No.2 was too close for comfort and suddenly he could not see what happened to the 190 or anything else.

> That was the most frightening thing that happened to me. I was shooting at this 190, and whether it was his vapour trail I went into, or whether it was my own, the whole cockpit iced up on the outside, and I was absolutely blind. If it ices up on the inside, you can scrape it off. I didn't know what to do. I climbed as high as I could go; I weaved; I couldn't see out. I flew home on instruments at 35,000 feet until the ice melted. It was very frightening.

He had no cameragun, and could only claim a damaged, but believes the 190 was destroyed. The rest of the squadron was engaged, and he went home alone.

Next day, 12 October, the squadron flew cover for motor torpedo boats bringing back a commando force from France. On 15 October it was back to Le Havre, with Circus 227. The bombing was good, but Gaze was just too late taking the squadron onto 20-plus 190s and 109s. Lorient was the target on the 21st, as B17s and B24s attacked the U-boat pens. Rather than approach directly from England, the bombers were routed out over the sea, in an attempt to surprise the fighter defences. However, the October weather was so bad that most of the bombers turned back and only the 97th Bomb Group pressed on to the target. Needing visibility for accurate bombing, the B17s went down to 17,000 feet through a break in the clouds. This did not please the escort fighters, and the whole formation was at a tactical disadvantage.

FW190s attacked the rear of the formation, shooting down three B17s and damaging six more. The remainder placed two-thirds of their 30 2,000-pound bombs within 1,000 yards (1km) of the aiming point. The massive U-boat pens could withstand such hits, but the buildings and equipment outside were more vulnerable. The Spitfires did not claim.

That afternoon, Tony's favourite and personal aircraft, 'G', was returned, and he tested it over the next two days. He was at Tangmere to see the Lancasters set off to bomb Milan. On 25 October, in SH-G, he led the squadron on Rodeo 102, to Dunkirk-St Omer-Gravelines, with everyone freezing at 30,000 feet. The President of Brazil presented two Spitfires on 29 October, in a ceremony at Hornchurch, and Tony was called on to be present. The month ended with two other uneventful patrols.

However, the September fiasco with 133 Squadron was sheeted home to him, and by 8 November he was back in 616

Squadron as B Flight Commander. He had been offered a chance to retain his rank of Squadron Leader by going to an OTU as CFI, but he refused. 616 still operated the Spitfire VI high altitude version from Westhampnett and Ibsley, and on the 8th Tony flew YQ-P on Circus 235 to Lille. Again he just missed jumping seven 190s, but later gave a Mayday call for a pilot in the sea, who was picked up but was dead. Other uneventful flights followed, with sorties to Cherbourg, convoy patrols, coastal patrols and other flights, but the Luftwaffe was not keen to engage. On 27 November, during a Rodeo inland from Dieppe, a goods train was strafed and it ran backwards downhill, providing some amusement for the Spitfire pilots. During the month, Gaze had reached his 1,000th flying hour and by the end of November had 327.5 hours on operations. On many November sorties he had flown Spitfire VI BS460 YQ-S.

December 1942 began with coastal patrols and aborted Rhubarbs, and on the 4th Gaze flew a Rodeo to Le Touquet-Le Crotoy. 'Bags of 190s seen inland; no scraps; a little flak near Calais.' It was his 100th offensive sortie.

But on 6 December it seemed that the recent lack of Luftwaffe reaction had been their way of conserving strength, as many 190s engaged. The target was Lille, and 616 provided rear cover to 66 Fortresses. Simultaneously, 100 Venturas and Mosquitos attacked the Phillips Radio factory at zero feet, losing 11 aircraft. Many 190s attacked the B17s, and Tony fired at one, saw no result and claimed it in his log book as 'very frightened'. On 11 December, Tony was sent on a reconnaissance to Le Havre, looking for an armed merchant raider. No ship was seen, but at a mere 3,000 feet he felt very exposed, as there was no cloud, and he was himself 'very frightened.'

The B17s set out across the Channel again on 12 December, but bad weather diverted them from their target at Romilly, and some turned back. 616 provided rear cover, and engaged the Focke Wulfs. Gaze fired at one which rolled inverted and went down in a long steady dive, trailing smoke, disappearing into the clouds below. He was sure that he had hit the pilot and that the aircraft continued its dive into the ground, so claimed a probable but as nothing was seen of a crash it was not credited to him. Gaze does not accept that any pilot would remain in an inverted dive, with smoke coming from the aircraft, down through cloud, with enemy above, even if for the simple reason that visibility when inverted was extremely limited. One possibility is that the pilot baled out unseen.

Bad weather again restricted flying operationally until 20 December, when US bombers attacked Romilly. The Luftwaffe was very well aware of the range of the Spitfire and prudently waited until the fighter escort had to turn back; then JG2 and JG26 attacked in relays to and from the target, ceasing as seven squadrons of Spitfires, tasked to cover the withdrawal, appeared on the return flight. Six bombers were lost, two more crash-landed on return and 29 others were damaged. The 8th Air Force accepted figures of 21 Germans destroyed and 31 probables, but Luftwaffe records indicate five were lost. 616 escorted some 50 B17s on their way out but had no combats although a few Germans were seen. The Spitfire VIs were taken to 33,000 feet on the return to England. On 22 December a Rodeo to Cherbourg was flown, with no combat, and the next day another was flown to Abbeville and 'places west'.

On Christmas Day several of the groundcrews were given local flights in the Puss Moth. The next sortie across the Channel was on 29 December, to Abbeville. Twelve Focke Wulfs were seen ahead of the Spitfires, dived away before they could be engaged but others were above, so the Spitfires did not follow them down. Next day, no enemy aircraft were seen at Cherbourg, but flak was accurate, and Tony described it as 'general shambles.' At Le Touquet next day, a 'fiasco' as the Germans were seen above and an 'undignified retreat' resulted.

January saw no operational flying for Tony, as his tour ended on the 8th. He was awarded a bar to his DFC and sent on a speech-making tour of aircraft plants which was intended to raise the morale and productivity of the workers. Originally he had been posted to 10 Group—'to my horror'—but he had made himself unpopular there by stating they were responsible for the disaster to the Eagle Squadron, so he did not remain long.

In other theatres of the war, turning points had been reached. General Montgomery won his battles at El Alamein, and by the end of the year the German-Italian forces in Africa were in retreat. Other Allied forces had landed in Morocco in November, creating a two-front theatre for the Axis. In the Solomons, the Americans had won the Guadalcanal campaign. The Japanese also had been defeated in New Guinea by Australian forces. The Russians had surrounded and annihilated the German 6th Army at Stalingrad. In Britain, the Communists, socialists and their sympathisers, ignoring the military practicalities, were demanding the invasion of Europe— 'Second Front Now!'—to assist their Soviet comrades.

From 6 February 1943 to 22 April, Tony Gaze was posted to PR3 Air Ministry. With other veterans of the three Services, he was sent to give pep talks to the factory workers. He was drilled by experienced speakers, and was surprised to have to sign a document stating that he would not engage in political discussions; he was required to obey several other restrictions on his speech and behaviour. According to the press and radio propaganda, the factories were working at maximum capacity but he had 'my eyes well and truly opened by what was going on.' The strikes and other union actions, regardless of the war and needs of the population and fighting services, came as a surprise. Tony did all that was asked of him, except for a 'Wings For Victory' event in Trafalgar Square. 'Shooting a line' to factory workers was one thing, but it was quite another to do it in Trafalgar Square, where there was certain to be someone present in the crowd who knew him.

From 22 April to 15 May, Gaze was in the Masonic Hospital for treatment of an eye infection—probably the result of dust or some other foreign body entering his eye whilst he was touring the factories.

> They couldn't find out what it was. They pulled a couple of teeth out; during the war they'd pull teeth out as soon as look at you, because you weren't going to live long, why patch you up? They were very worried about you having tooth problems, because as you climbed the problem became more painful, so you used to lose your teeth very rapidly. I ended up without a lot of teeth.

He then went as supernumary Flight Lieutenant to Hornchurch, and managed to fly operations with 453, 222 and 129 Squadrons until 20 June. Then he was a Fighter Tactics Instructor with 268 Squadron until 19 July, returning to strength of Station HQ Hornchurch. These attachments allowed him to fly operationally, as well providing opportunities to fly some of the current fighter aircraft such as the Typhoon, Mustang and Spitfire IX. Various types of offensive operations were flown—to Dieppe, Flushing, St Omer, Dunkirk, Schipol, Triqueville, Merville and Le Trait—with such log book comments as 'A few Huns dived away'; 'Bags of 109s and 190s'; 'inconclusive darts at milling mob of enemy aircraft'; 'mad shambles of Spits went miles inland'; 'bombers turned back'; 'many 109s and 190s, general mill around'. At Merville, on 31 July, 18 B26 Marauders bombed the dispersals, the Luftwaffe engaged, Tony got into position on a German but his air pressure was down and the guns would not fire.

On a flight with 453 Squadron RAAF, his engine would not develop power during the climb to height out of England, so Gaze had to drop the belly tank and look for somewhere to land. Below was a golf course, so he circled down to it, but at 1,500 feet saw that it was covered with a network of cables stretched some feet above the greens as an anti-invasion measure. Quickly, he picked an alternate — a small field — and put the Spitfire down safely. Then it became a race between himself and the converging groups of local children, as Tony tried to find a screwdriver to open the gunbays and unload the cannon and machineguns before the children could get into the cockpit and, the dream of every wartime kid, press the famous firing button on the Spitfire control column.

He got the screwdriver from a tractor nearby, and was also able to telephone 453 Squadron from a convenient pub, in which the local people insisted on buying him drinks, to the extent that when the recovery crew arrived, he 'could hardly move.' Tony was also mildly concerned that someone would remark on the squadron letters of this aircraft which had failed him. 453 Squadron used letters FU, and this particular aircraft was 'K'.

RAF Bomber Command created the first man-made firestorm in its attacks on Hamburg beginning 24 July, when the German defences were swamped by the new tactic of passing a large number of bombers over the target in a short time, and by dropping bales of metal-foil strips which reflected thousands of return signals to radar screens on the ground, negating control of the nightfighters. The fire-fighting services and air-raid defences were overwhelmed, and an estimated 40,000 people died.

On 4 August, while on the strength of Hornchurch station HQ, in DV-P, Gaze flew to Le Trait and attacked gaggles of 109s and 190s near Abbeville and Dieppe, but despite a 10-second burst at one, there were no results. He wrote up the German in his log book as 'frightened'. On the 15th and 16th, Flying Fortresses were escorted to Poix-Amiens and to Le Bourget, with no Luftwaffe interception. At Knocke on the afternoon of the 16th, Tony attacked a gaggle of 109s and 190s. He fired on a 190, but the 20mm cannon jammed and he pressed on with the machineguns; strikes flickered on the German, black smoke streamed out and the 190 went into a dive, going past the vertical, disappearing from sight.

17 August saw the beginning of the major 8th Air Force attack on Regensburg. Spitfires escorted the B17s to the limit of their range, but it was rare for the Luftwaffe to engage while the Allied fighters were about. However, on this day, Flying Officer Ray Hesslyn, a New Zealand ace from the Malta battles of 1942, saw a German formation estimated to be 23,000 feet below—Hesslyn had remarkable eyesight. Hesslyn and two other Spits dived to attack, but the remainder of 222 Squadron was left behind. Hesslyn shot down two, Flight Lieutenant Tripe got another and one damaged and the Messerschmitts scattered: III/JG3 had a tough introduction to the Western Front after transfer from Russia. After escorting the B17s as far as Antwerp, the Spitfires of 129 Squadron turned back, but also saw enemy below over the Scheldt, and dived from 28,000 to 4,000 feet. In the dive, Tony Gaze passed both the squadron leader and wing leader, and went after a FW190, fired a six second burst from 300 yards (metres) to 150, strikes flashed all over the 190 and the pilot baled out. However, it was Spitfire 'A' again, and the cannon had jammed once more. The German unit cannot be identified, nor the loss traced, but the victory was confirmed by other

pilots and his cameragun clearly showed the FW190, and the strikes.

In the afternoon further missions were mounted to meet the returning bombers, who received a severe hammering as soon as the fighter escort turned back. The Luftwaffe had attacked the bombers all the way to and from the target. After such a battering, by the time the formations reached the welcoming RAF and USAAF fighters on the return flight, there were aircraft all over the sky, but the Germans had relented, and no enemy were identified by Gaze's squadron. During the day, US Thunderbolts shot down and killed Major Wilhelm-Ferdinand 'Wutz' Galland, brother of the famous Luftwaffe fighter General. Major Galland claimed 55 victories; another fighter pilot brother, Paul, had already been killed and a fourth, Victor, was captured in England.

Next day, 'A' lost its belly tank on take-off, so Gaze quickly returned and landed, scrambling in 'E', caught up with the squadrons and found that they then turned back because of cloud. Later in the day he scrambled to shoot down a stray balloon at 23,000 feet, and 'A' had another cannon jam. Confidence tends to fall off after three failures in three days, and this was the last time he flew the aircraft. Next day he flew Spit IXB DV-? to Poix as escort for B26 Marauders and whilst there was no contact the flight gave him an opportunity to assess the IXB, as 'pretty good'.

Later on 19 August he flew it again, escorting Fortresses to Brussels. This strike was made largely by crews who had not flown the Regensburg-Schweinfurt mission of two days earlier, but weather prevented accurate bombing or good results. Tony attacked a Bf109, which was fitted with the prominent under-wing launcher for the 21cm mortar shells intended to break up US bomber formations. Gaze fired from 350 yards (metres) to 150 following the 109 in a dive with 530mph (830kmph) on the air speed indicator, but he could not see any hits as his windscreen became covered with oil and so he broke away. He did not see any results so did not claim. Later the camera film showed strikes and the Intelligence people told him that the tail had been shot away. But as he had not claimed it before, they would only allow him to claim a probable. The attitude of the Intelligence staff was that if he had not claimed it initially, then no claim would be allowed, despite what the film showed . . . German records show Oberleutnant Johannes Meyer, 10 Staffel III/JG26, crashed in his aircraft after being badly damaged by a British fighter.

The recent sorties had been flown with 129 Squadron, but on 26 August Gaze was posted to 66 Squadron, Kenley, as A Flight commander. 66 had relatively old Spitfire VBs, not as happy a situation as flying IXBs at Hornchurch. The VB was the original 'clipped, cropped, clapped' aircraft, and by mid-1943 was obsolete. Next day, after joining 66, Gaze flew to Beaumont-le Roger, but the Boston bombers turned back because of thick cloud. In the afternoon, St Omer was attacked by 240 US heavy bombers, and the squadron was part of the close escort to the last box of 60 B17s. Flak was accurate, and FW190s attacked but one was destroyed. The bomber crews had been told that they were to attack 'special aeronautical facilities', and to assist in accuracy the bombing height was only 14,000 feet. The targets were part of the German V-weapon launching network, and the attacks had been made when the concrete was not hardened. The bomber crews were pleased at the short trip, coming after the long hard slogging matches across Europe in recent weeks. Many more short range attacks would be made in the 'Crossbow' campaign against the V-weapons.

During August, the Allies had advanced in the Pacific and in Russia, while in the Mediterranean preparations were almost complete for the crossing from Sicily to Italy. On 17 August,

RAF Bomber Command attacked the German rocket development site at Peenemunde, their targets being the accommodation blocks and administration buildings occupied by the German scientists and specialists. Many were killed, and some of the survivors were resentful that they should be bombed, ignoring the fact that their work was intended to enable Germany to bombard cities indiscriminately.

On the morning of 4 September, in LZ-A, Tony flew as part of the escort to Bostons bombing Amiens; he had 1140 hours flying time. Alone, he got into a fight with a staffel of FW190s, at low altitude, shot one down, but followed it too far down and the others came after him. In the combat, he hit another which flicked inverted at 500 feet but was not seen to crash. Tony later found that the 190 would flick if the pilot overdid his use of the electic trim to assist in manoeuvring, and he still wonders if this is what happened on this occasion. JG26 lost two pilots this day near Amiens: Leutnant Ernst Heinemann, 4 Staffel II Gruppe, with two victories claimed, and Oberfeldwebel Walter Grunlinger, who had 7 victories, flew with Stab/JG26. Grunlinger was wingman of the ace Joseph 'Pips' Priller. Luftflotte 3, the German formation in France, recorded a loss of three FW190s destroyed and one missing on this date.[1]

Gaze called to the rest of the squadron, 'I need a bit of help here; I've got six 190s', but no one else arrived. With the pack of 190s after him, Tony used their shadows on the ground to keep aware of their relative position, and when the Germans were 500 yards (metres) behind him, he would whip around and face them. He continued fighting after he knew he could not get back to England. He saw, in the shadow cast by the Spitfire, 'a great trail coming out, and knew it wasn't long (to go) and I'd have to park the thing somewhere pretty quick.' He called on the radio, 'Don't bother, I've had it.' The

Spitfire was streaming glycol and petrol was down to twenty gallons (90l). He had been hit in the elevators at the start of the combat anyway, and crash-landed behind Le Treport. The Spitfire was claimed by Gerhard Vogt, II/JG26. Vogt rose from Obergefreiter to Oberleutnant and command of a Staffel, was awarded the Knight's Cross in November 1944, and killed in January 1945, credited with 48 victories.

The rest of the squadron, not having seen Tony, but only hearing the radio messages, thought he had been killed, and he was reported 'Missing, believed killed'; his family was

Gerhard Vogt, JG26, shot down Tony Gaze near Amiens on 4 September, 1942. Vogt, with 48 victories, was killed on 14 January, 1945. Gaze was Vogt's 14th victory.

notified. 'It was over-confidence and frustration,' Gaze recalled, '(after) having the Spit Nines dive in front of you, and every time you went to have a go at anything . . . I followed this thing down too far and got in amongst all his friends. It was inevitable. I had no fuel, I was trying to make for the coast, instead of inland . . . Stupid. I don't know if I thought I could paddle my dinghy over the sea.' In what he later realised might have been good fortune, in the crash-landing he damaged his face on the gun-sight. Had he been unhurt, he believes he would have tried to do something 'like rush off and try to steal a 190 or something equally silly.'

As it was, he was taken to a loft by Resistance people who reached him before the German search parties. The Frenchmen were what Tony described as 'gloating', quite pleased after having seen and examined the crash of a FW190, in which the pilot had been killed. Tony determined that this was not the 190 he had shot down, as that one was too far away, and wondered if it was the one which had flicked onto its back at low level during the combat. He waited while his face healed somewhat, and then the Resistance arranged for him to be taken to a field near Arras where a Lysander was to pick him up and fly him to England. In typical Gallic fashion, his departure was reason enough for a small party—discreet, but a party nonetheless. There were drinks and music, which included a one-finger piano rendition of 'God Save The King' played while Germans were actually outside the building. Tony found it somewhat amazing that no one seemed to pay the slightest attention to the singing, though it was in English. He decided it was more a comment on the ability of the singers.

It was some twenty miles (32km) to Arras, and the distance was to be covered by bicycling. Two bikes were produced, and Tony set out, following his guide, who pedalled along some distance ahead. They came upon a group of German officer cadets who were being trained in map-reading. With them was an anti-aircraft gun, barrel elevated, crew scanning the skies. The Frenchman almost fell off his bike, laughing at the sight, and though the Germans were not amused, there was nothing they could do about it, and both cyclists went past safely. Tony caught up with the Frenchman, and asked what caused the hilarity. The reply was that it was due to the sight of German gunners peering into the sky for a British pilot, while one was pedalling past a few metres away...

But the Germans had broken that Resistance circle, were arresting its members and, having learned of the planned aircraft visit, were awaiting it at the landing ground. Luckily, one woman had so far remained free and warned the group bringing Gaze to the landing place. They hid in the bush for the night and, as no one knew what to do with Gaze, they then sent him to Amiens. From there the Resistance sent him to Paris, 'with the usual rules — come out of the station, look for a chap in a white hat with a cigarette and follow him.' This was done, and Gaze shared a flat with a French Canadian. Having the supply of escape money given aircrew, they were able to survive easily enough by eating in the black market restaurants, which did not require food coupons. After a time he was offered a choice of waiting until the next moon period and a night flight from a field to England, or going south and crossing the Pyrenees, said to take only four days.

Being keen to get moving, Gaze opted for the mountains, and found that instead of four days, it took ten days of climbing through and over them. Tony's journey through France was remarkably uneventful, except for one incident at Toulouse, when the Italian guards became suspicious, and Gaze and the guide had to run alongside a departing train to the station entrance, then outside to a restaurant, where they sat and waited to see if anyone was following them. Tony was surprised to see that the organisation even included a 'girlfriend' for each escaper. The couples would stroll out of town together, as though unwarlike activities were all that was on their mind, and meet the guides in the countryside. The girls would return to town, and the party would set off into the mountains. The mountain guides were Spanish smugglers, who had found it more profitable to bring people rather than contraband through the Pyrenees.

Halfway through the journey, the party met a group of USAAF escapers who had been abandoned by their guides. The reason for this soon became clear. The Americans refused to accept any orders from the guides except those they personally wanted to obey, and ignored basic safety requirements, such as not smoking openly at night. At halting places, out on the open slopes, the guides would come along the line of men, telling them not to smoke and not to make noise. The American reaction was that 'no goddam Spaniard is going to tell me what to do,' and to defiantly light a cigarette. These men ignored the fact that the guides would be shot if captured, while the Allied servicemen would only be sent to a prisoner-of-war camp.

Tony was surprised at the attitude of these Americans to the war. Their view was that, having been shot down, if they escaped successfully, they would be returned to the USA and not be required to leave it again; they knew which medals they would receive on return to the US forces; none wanted more operational service. His own comment, that he wanted to get back to a squadron, in hopes of finding the German who had shot him down and was the cause of the exhausting mountain climbing, was not echoed by any similar spirit of aggression in the USAAF group. When the party finally came out of the mountains, the Americans were again disillusioned. There was no one to meet them—no cars, so they would have to walk to Barcelona, which they did with constant complaints.

Tony Gaze DFC

When he arrived at the British offices in Barcelona, Tony found that the cellars contained a veritable Moss Bros., but he was too tall for the available selection of clothes, and had to be outfitted by a local tailor. He found that the Spanish assistants were helping the British because they thought the Allies would eventually win the war, whereas some of the French were doing so because they were Communists and had been told to do so by the Party. In Gaze's opinion, formed after his time amongst them, many of the French resistance were patriots but many others were criminals whose activities suddenly became legitimate.

At Madrid there was what amounted to a transit camp for the escapees, though everyone was given a false name, which was on a list periodically given to the Spanish authorities, who would then deport those on the list. Tony was given the rank of Major, and a fictitious name, for the purpose. However, on arrival at Gibraltar, he forgot the new name, and the Spanish official had to call it several times. But there was no problem, and all the group finally crossed into British territory.

Tony Gaze was returned to England on 28 October 1943 but, on arrival, was a little irritated to be interrogated for three days 'as though I was a crook'. He was admitted to Halton Hospital for treatment to wounds suffered when his face hit the gun-sight during the crash-landing. Offialdom did not want him to go back on operations, and he was medically graded unfit for operations for six months. He went to the Fighter Development Unit at Wittering, then to the Fighter Affiliation Unit at Swinderby in February, and experienced great difficulty in getting back to operational flying. Official policy was that escapees should not be returned to operational flying over Occupied Europe, in case they were again shot down and captured, then forced to reveal details of the people and system which had initially returned them successfully to the UK. For the next four months he flew and fretted, using old friends in the active fighter Wings to allow him to sneak a few operational sorties. Then came D-Day, and the Normandy invasion.

In the meantime, the Allied armies had landed in Italy and begun the painful series of attacks north through the mountains and river valleys which favoured the defender. By the end of May 1944, Rome was about to fall to the Allies; MacArthur's forces had almost isolated New Guinea, leaving the Japanese forces there cut off from supplies or rescue; there was fierce fighting in Burma. Of great concern to aircrew was the announcement on 19 May that 47 RAF officers who had escaped from Stalag Luft III had been shot on recapture. This was against the Geneva conventions and became another war crime held to the account of the German leaders. One of those shot was Gordon Brettell, who led 133 Squadron on the escort mission in September 1942 when they dived into the air defences of Cherbourg.

On 5 June, Tony returned from a training flight and noticed the groundcrews busily painting black and white stripes around the fuselages and wings of aircraft. He realised at once that this was a sign the invasion was imminent. When the great operation did take place, Gaze was airborne among the thousands of US and RAF aircraft of all types flying to France, and a curious observer of all that was going on until he realised that of all the Allied aircraft parading across the Channel and along the Normandy beaches, he was alone in not having the distinctive black and white stripes. He then smartly removed himself from the area.

The squadrons expected hordes of Luftwaffe aircraft to react to the landing. Unwilling to be out of the expected upsurge in fighting, Tony had absented himself from Swinderby, and attached himself to Johnny Johnson's Wing of Canadian Spitfires. Several sweeps were flown with no significant action, while the ground forces fought their way inland, and behind them engineers hacked airfields from French fields. The Wing was to take off from Ford, fly a sweep, land and refuel in France on the newly-captured and prepared airfield at 'B-2', Crepon Bazenville.

Tony was to fly with them as a supernumery unofficial member of the formation. There was a certain amount of media reaction and arrangements to meet the first Canadian fighter pilots to return to the Continent. As the Wing was taking off, one of the Canadians burst a tyre, and Johnson called Tony on the radio, telling him to wait and then bring the other pilot along when the wheel had been changed. So Tony, with the Canadian as wingman, patrolled the beaches but no Germans were seen. 'We were expecting to get into a hell of a battle,' said Tony. 'We thought there'd be Germans everywhere. In fact, I thought that with only two of us, it was a bit rash. We saw lots of fighting going on underneath us, but didn't see any Germans at all.'

Below were the dusty summer fields of Normandy, and Tony anticipated landing in a cloud of dirt and grit after the 36 Canadian Spitfires had landed. He then called his No. 2 and dived ahead of the three squadrons, landed and taxied to the far end. There was quite a crowd gathered, and before he got away from Spitfire 9G-R the media representatives were clamouring for his name and place of origin in Canada.

The reply that he was 'Tony Gaze, from Melbourne, Australia,' put quite a damper on the enthusiasm. 'I had to hide, though,' he recalled, 'as Broadhurst came, and he knew me.'

Tony Gaze therefore claims that he flew the first Spitfire of the first Wing to land and refuel in France after D-Day. A couple of others had been sent by Johnson to check on the arrangements and suitability of the field for Spitfires, but they were not part of a Wing operation.

After a few days, with no combats, Gaze returned to

Swinderby to a quiet reception by the Station Commander who required him to explain his absence. Tony said that he had been in the war from the start, and was not going to miss D-Day. The Group Captain informed him that his punishment would be announced over the Tannoy loudspeaker system. He was to talk in the Mess for half an hour on a nominated subject without any preparation. This was done, and the time limit was enforced. Tony regards the punishment as reasonable. The Group Captain could have applied relevant King's Regulations, and both realised that if Tony had asked permission to go to the invasion, the answer would have had to have been negative.

For the remainder of June 1944 Gaze had to content himself with fighter affiliation flights. These continued into July until, on the 22nd, he was posted to 610 Squadron at Friston as a deputy Flight Commander. He had been packed and about to leave on a conversion course to Mosquitos, for a tour of marking targets in Path Finder Force, when suddenly the posting arrived. The squadron was equipped with Spitfire XIVs, having the Rolls-Royce Griffon 65 engine and five-bladed propellor married to the Spitfire VIII airframe. The squadron had re-equipped with the MkXIV on 1 January and was to be used against the V-1 flying bombs launched against England after Normandy was invaded. Tony's comment after flying a Spit XIV for the first time was 'good kite but over-powered. Terrific climb.'

As he was arriving at the airfield, Gaze was informed that a Beechcraft was in the area. He acknowledged and looked around but could not see this US-built light aircraft, so continued and landed. There was a certain amount of excitement, with people asking if he had seen it, and then he was told that 'Beechcraft' was the code-name for a V-1. 'Diver' was the usual code-name for them, and he was a little chagrined at the lost opportunity to get one on his arrival flight.

Gaze regarded the anti-V-1 patrols as 'very frightening', entailing flying unreliable machines, sitting five miles (8km) off the French coast with a rough engine and being told to wait until relieved. When a V-1 appeared, the Spitfire XIV would chase it across the Channel and, as it was almost in firing range, the V-1 would enter the anti-aircraft barrage zone, when the guns would fire regardless of the RAF fighters. If the V-1 survived the barrage it emerged into the patrol areas of the Tempests, which dived on it. The muttering Spitfire pilot turned away, back over the Channel.

For the rest of July he flew patrols waiting for the V-1, only able to shoot a few pieces off one which had already been hit hard by a Tempest and was going down. Even the exploding V-1 was dangerous as the blast could destroy the pursuing fighter, and he noticed that the fuse was so sensitive that the missiles exploded immediately on contact with the sea. There was no splash or pillar of spray, just a 'great explosion that nearly shook the wings off.'

By the end of July, his operational hours totalled 410.35, and he continued patrols against the fast small Divers.

On 5 August, a V-1 came sparking along from behind, and Gaze was able to time the turn to bring the Spitfire XIV in behind, in range, to fire, and destroy it. It was his sole victory over the pilotless machines. Next day he chased another for 15 miles (25 km) but he could not catch it. On 14 August, in DW-T, the oil pressure collapsed and he had to return, sweating out the flight, with 'bags of panic!' Patrols continued with no contact, but on the 19th, as soon as he had landed, many V-1s came blatting across the sky, while on the 23rd many had passed just before the Spitfires arrived on station.

In July the US and British armies broke out of the Normandy bridgehead and surged across France. Inept plotters tried to kill Hitler with a bomb but failed, and the Fuhrer launched a wave of arrests and killings in retaliation. Southern France was invaded on 15 August. The Germans had suffered heavy losses in Normandy. Rommel had been wounded by air attack, and Field Marshal Kluge committed suicide rather than go back to face interrogation about his part in the plot to kill Hitler. Paris was liberated on 25 August, and five days later the Russians occupied the Rumanian oil fields at Ploesti. In July and August, the German air formation defending France, Luftflotte 3, had lost 1,554 aircraft to operational causes.[2]

By September it seemed that the worst of the V-1 offensive was past and 610 Squadron, together with 41 and 350 Squadrons, began escorting bomber formations from the UK to European targets. On 10 September, Marauders dropping leaflets on German forces trapped north of Ostend were shepherded through accurate flak but no enemy aircraft were seen. Next day, 200 Lancasters attacked oil targets at Gelsenkirchen, again through flak, but no fighter attacks were made by the Luftwaffe. On 12 September a similar escort was flown for Halifaxes bombing oil targets in the Ruhr. That day, US forces crossed the German border near Aachen. On 13 September, Gaze patrolled the Hague area on an anti-V-2 patrol, then later escorted B25 Mitchells bombing the viaduct between islands at Schouwen. Next day, an armed reconnaissance was flown in the area south of Amsterdam, with attacks on barges, wharves and warehouses, but the Griffon powering DW-T cut, resulting in 'much panic'. The War Diary of Luftflotte 3 referred to the fighter-bomber attacks, estimating that 310 aircraft took part, 'with considerable nuisance effect in places.'[3]

On 17 September, Operation Market Garden began. This was the ambitious attempt to unroll a carpet of Allied airborne forces, from the British front line, across Holland to the Rhine, thus opening up the northern plains of Germany to

Montgomery's divisions, and perhaps ending the war before Christmas 1944, while the Soviets were still in Poland. 610 was tasked to attack any flak positions which opened fire on the procession of gliders, tugs and paratroop-carrying aircraft. Tony noted that there was little flak, and all the people in the villages came out to wave. The Spitfires hammered a flak-ship, leaving it on fire.

On 19 September, a planned mission patrolling the Market Garden area was cancelled, after it was airborne, due to poor weather. The squadron had to attempt to land in heavy rain, and the first three crashed, the remainder going on to Manston. Tony, landing with a full 90 gallon (409l) drop tank, found himself overshooting and pulled up the wheels to avoid hitting the aircraft which had landed in front of him; its undercarriage had collapsed. With 90 gallons underneath his fuselage, it was 'very fraught'. The enquiry into the incident found in his favour, 'why I can't guess'.

Next day an anti-flak patrol was flown during a re-supply mission to the Arnhem area. There was no Luftwaffe activity but the many Allied aircraft on the ground were very visible, with the white and black stripes on wings and fuselages. In addition, in one of the towns below, a Liberation celebration was in progress so the Spitfires provided a fortuituous and timely 'beat up'.

On 23 September, Dakotas and Stirlings were escorted to Arnhem but there was no Luftwaffe activity, though the low, slow dropping formations suffered terribly from ground fire. It was obvious to the Germans where these aircraft were headed, and their passage was radioed by garrisons holding out on the coast, allowing plenty of time to concentrate maximum firepower along the approaches to the dropping zones. During the mission, the squadron crossed the Seigfried Line, going into Germany for the first time. A similar sortie was flown on the 26th. There were 'thousands of kites milling around, all types.' One enemy jet was seen by Colin Gray, but there was no other activity. This mission was 2 hours 50 minutes, 'too bloody long!'

On 27 September, the Spitfires landed at Antwerp. It started to rain, so the pilots sheltered under the wings of the aircraft. Then some shelling began, and they realised that the Germans were aiming at the fighters, so there was an exodus of pilots away from the Spitfires. Every time a shell came in, they threw themselves flat in the rain, mindless of the muddy grass, so that soon they were quite dishevelled. The Germans were only 3,000 yards (metres) away, but the closest shell was some 38 feet (12m) from any of the fighters.

The pilots decided to explore the city and soon found that their untidy uniforms proclaimed to the citizens that here were real combatants. Other RAF personnel stationed in the city had been ordered to wear their neat and clean best uniforms, in order to make a good impression, but were almost ignored by the population. To the annoyance of the more respectably dressed RAF, Gaze and his companions were the recipients of boundless hospitality, and what followed was a 'pissy evening' at the Hotel Century and other locations.

In July, August and September, Luftflotte Reich, the German air formation defending Germany itself, had lost a total of 1,645 aircraft to operational causes. Throughout September, despite an increasing weight of aerial attack, the German aircraft industry actually reached its peak production for the entire war, when 2,876 new fighters rolled out of the factories. More importantly, Luftwaffe losses for the month were 371, for 307 Allied aircraft claimed. During the month, the Russians advanced to the Baltic, into Yugoslavia and further into Rumania. In Italy, the Allied armies were also moving forward. In the United Kingdom, demobilisation plans were announced. For the aircrews and front line soldiers the war went on in their immediate vicinity, just as before.[4]

During October, the Germans resisted fiercely, proving that they were still a dangerous enemy. On the Russian Front, in a 19-day battle around Debrecen, three Soviet Corps were destroyed. V-2s began to fall around Antwerp and in Hungary Otto Skorzeny forced Admiral Horthy to continue the war on the German side. Germans aged between 16 and 60 were enrolled into the *Volkssturm*, and at Auschwitz the last of the extermination gassings took place. November was to be another month in which the Germans fought on—unconditional surrender being the only alternative—ruled by a merciless government using the stick of horrors to be inflicted on Germany by their uncivilised foes and the carrot of 'wonder weapons' soon to be employed.

Tony flew bomber escorts with no fighter combats, though bombers were lost to flak. The Spitfires could not reach deep into Germany, where the Luftwaffe was concentrated. Pushing eastwards into Germany, the US 8th Air Force was fighting large battles with defending fighters, and Luftwaffe figures show that 120 fighters were lost on 2 November, for 40 four-engined bombers claimed shot down. 180 RAF Lancasters bombed the Ruhr on 4 November, with Tony Gaze flying as one of the escort on Ramrod 1359, followed by another on the 9th, summed up as 'usual 20,000 (feet) dice'.

On 14 November Gaze flew base-Amiens-Glisy-Juvincourt-base as part of the escort for Prime Minister Winston Churchill's visit and return flight to Northolt. During it he was able to find the site of his crash in September 1943. On 16 November, the squadron escorted Path Finder Force Mosquitos at 30,000 feet, as they marked targets for 1,150 Lancasters and Halifaxes attacking German areas in front of the US 1st Army. Then, on the 21st, Lancasters bombed Hamburg for the first time in daylight and the squadron escorted them. More than 30 bandits were reported but not seen, though Tony did sight four V-2 launches, vertical trails of smoke and propellant spearing up into the stratosphere—useless signs of German technological advances in the face of Allied power. There was heavy flak and a 'scarecrow'—what RAF crews believed to be a German shell which exploded and simulated a destroyed bomber, supposedly to intimidate them. Experienced crews could distinguish between a 'scarecrow' and the real thing. But the Germans never did use 'scarecrows': The explosions were real bombers.

Despite worsening conditions for ground fighting, the Allies continued their advances, and were inside Germany in many places. On 7 November, President Roosevelt was re-elected for a fourth term, with Harry Truman as Vice-President. On 12 November, 617 Squadron sank the German battleship *Tirpitz* in Tromso Fjord. On 21 November, the German fighter arm lost 62 aircraft, but only four heavy bombers and two fighters were destroyed in return. On the 26th, another 87 fighters were lost, with 20 bombers claimed shot down, and next day 51 fighters were lost, against a total of 11 Mustangs. By the end of the month, I Jagdkorps had lost 404 fighters, for 155 Allied aircraft destroyed. In the two months October-November, Luftflotte Reich had lost, by their own records, 733 aircraft to operational causes.[5]

On 30 November, Tony flew again as escort to 60 Lancasters bombing synthetic oil plants, through solid cloud, and again there was heavy flak and a 'scarecrow' but no fighter opposition. The pressure of attacks by day and night on Germany was maintained, and on 5 December Luftwaffe records show a loss of 75 fighters, with only five US bombers destroyed.

On 15 December, Tony took over B Flight 610 Squadron.

Next day the Germans launched the Ardennes Offensive, and the American forces staggered back under the impact of a carefully planned and executed assault. It was the greatest defeat in their military history, up to that time, and entire units disappeared in the chaos rolling westwards under the winter weather. However, US General George S. Patton had not been taken by surprise, and in one of the best changes of direction by a major formation in modern warfare, he swung his 3rd Army divisions north, across and behind the American front, to attack the southern edge of the German penetration. Field Marshal Montgomery moved British forces from the north, and assumed overall tactical control. Montgomery was to command all Allied forces north of the salient, with General Omar Bradley commanding all those to the south. Unable to resist rubbing American noses in the dirt of their predicament, Monty went about holding the Germans while enraging his Allies.

For Tony Gaze, this tactic meant flying in support of the US forces, who were quick on the trigger and shot at anything not carrying their own national marking—and sometimes even at them. 610 patrolled Aachen-Malmedy, and were 'shot at by Mustangs', 'shot at by Yank flak', 'shot at by Hun flak', and 'shot at by Yank flak'. For the Spitfires, and other aircraft, the support missions meant, as Tony recalled, 'chasing around over white-coloured snow looking for white-coloured trucks and shot at by guns covered in white. All very fraught.'

On 27 December, Lancasters were escorted to bomb Rheydt, then on the 31st during another sweep the Spitfires were again attacked by US fighters. Fortunately the superior rate of climb of the Spitfire XIV enabled the squadron to evade the lunges, then spiral up out of reach of the Mustangs, while the pilots looked down to see the P51 wing tanks jettisoned as the American pilots unsuccessfully tried to keep up.

In the afternoon, 610 left Evere for 'Y-32' airfield, near Hasselt. On the other side of the lines, the Luftwaffe was preparing for a great blow against the Allied airfields in northern Europe, an attempt to throw their opponents off balance and perhaps buy enough time to redress the situation in Germany's favour.

Allied fighter sweeps and escort operations had continued to reduce the available strength of the German fighter arm, which lost 50 on 27 December, another 31 to RAF fighters on the 29th, and a further 40 on the 31st. In the last four days of the month, 128 fighters had been lost, for only 86 claims, while Luftwaffe pilot losses in the final week of 1944 had been 316 killed, wounded or missing. Throughout December, attacks had been made on about 80 German airfields, destroying 129 aircraft on the ground and damaging another 140. These are German figures, not Allied claims. Goering had sown the wind with his Luftwaffe in 1939, and by 1944 the German people were reaping the whirlwind. In bombing attacks in Germany in December 1944 alone, 15,025 people were killed, 16,900 wounded, 242,845 made homeless and 65,000 homes were destroyed or damaged.[6]

All over north-western Germany, in the early morning of 1 January 1945, Luftwaffe fighter pilots prepared to take-off for a series of massed attacks on Allied airfields: Operation *Bodenplatte*. Led by very experienced commanders, the Messerschmitts and Focke Wulfs, sometimes guided by Junkers navigational aircraft, set off across the winter landscape. The first hitch came when their own flak opened fire—it was a long time since such massive formations passing overhead had been German! Then navigational errors were made, some targets were hard to find in the morning mist and, in places, Allied fighters were airborne and attacked. Nevertheless, many attacks came as a surprise and varying amounts of damage were inflicted on US and British squadrons. 126 Wing, composed of 41, 130, 350 and 610 Squadrons, was attacked by I, II and III/JG11—in all, some 60 Messerschmitts and Focke

Wulfs which flew from bases south of Frankfurt. Their attacks destroyed or damaged ten Spitfires of 130 and 350 Squadrons; 41 was away on an armed reconnaissance over the Bonn-Coblenz area.

Nearby, US Mustangs of the 352nd Fighter Group were actually taking off, and Lieutenant Colonel John C. Meyer shot down one FW190 as his P51 left the ground. Soon, P51s and P47s were chasing Messerschmitts and Focke Wulfs at tree top level around the airfields and through all this some Spitfires also managed to take off.

Tony Gaze got off in DW-T, but became separated from the other Spitfires and then, just east of Asch, saw and attacked a long-nosed Focke Wulf FW190D—the last one in a flight of eight. These were probably a staffel of Hauptmann Siegfried Lemke's III Gruppe JG2, which had attacked the airfield at St Trond, then turned back north-east for home. Gaze slid into position behind the Germans, and fired twice at the 190D without effect, but as the 190s lifted to clear a patch of forest he fired again, seeing his cannon strike against the belly of the plane ahead; its flaps went down and the 190D crash-landed at high speed, then burned. Luftwaffe records indicate that this victory may have been over Unteroffizier Werner Hilbert, 9 Staffel III/JG2, who crashed near Bad-Nauheim. However, other pilots in the flight reported being attacked by two Mustangs.

The Mustangs and Thunderbolts did arrive just then, and tried to shoot down everything not American. Concentrating on high-speed flight at low-level, the other members of Hilbert's flight may not have seen Gaze's Spitfire, but did see the Mustangs which attacked just afterwards. The Spits had to climb above, but saw the Americans shoot down a Typhoon which had entered the fray. Friendly and enemy flak were blazing away at anything which flew and which could not be immediately identified and the day was 'quite fraught' as Tony wrote, in his usual understatement.

Allied losses were some 152 aircraft of all types destroyed and 163 damaged. Luftwaffe records show losses of about 220 pilots and about 300 aircraft in the attacks. Worse than simple losses were the casualties among the more experienced pilots. 136 wrecked aircraft were counted in Allied lines, and some 40 others are still missing. In addition to the losses from *Bodenplatte*, elsewhere that day 27 Luftwaffe fighters were lost, for only 16 claims. It took some time for each side to discover how well or badly it had fared, and the usual conflicting claims for success and destruction were made. In some Allied quarters there was surprise and shock that the Germans could launch such an attack, but the material losses could be made good with little problem. There was no real lessening of Allied air pressure on the Germans, as the weather was often quite bad and sorties could not be flown, while replacement aircraft were brought up.

On 5 January, two close escort missions were flown for B25 Mitchells bombing targets at St Vith. Then, Tony had a break until the 13th, when the sole sortie by him was a long 2 hour 15 minute mission escorting more B25s, after which he went off on armed reconnaissance. On 14 January, there was a scramble for the Houffalize area, where Germans were reported but the Spitfires were recalled, then flew an armed reconnaissance to St Vith, during which some trucks were found and strafed, despite intense light flak. The Luftwaffe had been airborne that day, against US bomber formations, and losses were recorded of 139 aircraft, 69 from JG300 and JG301, with 107 pilots killed or wounded. One of those killed was Oberleutant Gerhard Vogt, Staffelkapitan 5/JG26, who shot down Tony Gaze near Amiens in September 1943. Tony Gaze had been Vogt's 14th claim, and the pressure of the airwar can be seen from Vogt's victory list: two claims in November 41, nine in 1942, six in 1943, but 31 claims in 1944. Thirty three were fighters.

On the 16th, another armed reconnaissance to the area south of St Vith drew a fierce reaction from all types of German flak, but no ground targets were seen. However, DW-T had its starboard wing riddled by an 88mm flak burst. Tony nursed it back to base and said good-bye as it was deemed category A/C damage.

Then from 19 to 29 January, the squadron's tired pilots were sent skiing at Megeve in the Haute Savoie. 'The Luftwaffe had just moved out,' said Tony, 'and the RAF moved in. This was the only sensible rest I had in the RAF, instead of sending people to be instructors, which they were not, or lecturing round factories of which they had no experience in doing.'

On return from the invigorating leave on skis, Tony flew several sorties in DW-S—armed reconnaissance in the Lingen-Rheine-Munster area, despite much light flak, attacking trucks, farmhouses near them, and trains found puffing across the snowy landscape. DW-T was back on 6 February, and more sorties were flown, one escorting Air Marshal Tedder in his Fieseler Storch to B56, Brussels-Evere, and attacking barges and trains in the Osnabruck-Paderborn area, plus two patrols in the Venlo-Nijmegen area, as the Luftwaffe was active there.

It was a case of 'third time lucky'. On the afternoon of 14 February, in the Nijmegen area, Arado Ar234 jet bombers, with Me262 escort, attacked the bridges. Tony led a section of two Spitfires on patrol and engaged the fast German bombers, who got away. Tony shot at two Arados, without result, then told his No. 2 to stay below cloud while he went above to look for enemy. As he broke through into the clear, there were three Messerschmitt 262s. He selected the starboard jet, fired and hit it, setting the starboard engine on fire, and the Messerschmitt rolled into a dive to the ground near Emmerich, killing the pilot, Warrant Officer (Feldwebel) Hoffmann.

This Me262, 9K+NL, numbered 110615, of I/KG51, was the first jet confirmed destroyed for 125 Wing aircraft.

On 16 February, Tony flew cover for medium bombers attacking Goch, then an armed reconnaissance to Rheine; 'Many weird weapons encountered.' Another similar mission was flown on the 21st, then the squadron returned to Warmwell in the UK for bombing and gunnery practice. Suddenly they were told that 610 Squadron was to disband.

Hundreds of RAF and USAAF four-engined bombers continued raids against German cities. Thousands of people died in terror-filled days and nights as their leaders refused to surrender, and the police dealt with anyone who dared say anything that meant, or could be thought to mean, Germany would lose the war. On 3 February, Berlin had been heavily attacked from 11.02 to 11.50; the bombers concentrated on the inner city, and killed 516 people, wounded 857, leaving another 1,500 missing and 100,000 homeless. On the 26th another 471 were killed, 917 wounded, an estimated 600 were still under the rubble as the report by the authorities was written, and 102,000 were homeless. In another raid on 18 March, 227 died, 849 were wounded, 450 were missing and another 65,000 were made homeless. Berlin was only one target in the shrinking Reich. This attack was by 1,200 heavy bombers, and from the endless formations the Luftwaffe managed to shoot down 24 bombers and five fighters. Another 40 Luftwaffe pilots were killed or wounded on 24 March. Desperately trying to mount a credible defence, the Luftwaffe had been preparing for another great blow against the daylight bomber formations, and on 7 April committed some 120 to 180 fighters to intercept. It was a disaster for them, as only eight bombers could be claimed, but 133 fighters were destroyed by the US escort or bomber guns. Massed squadrons of jet fighters, with reliable engines, flown by experienced pilots, were all that could save the towns and cities of Germany from the hammer blows delivered by the RAF and USAAF. But the fighters were too few, engines were unreliable and experienced pilots daily diminished. Hordes of Allied fighters rushed to every location where jets were reported, and waited to pounce when the jet was landing or taking off.

Faced with the prospect of more fighter affiliation exercises, or similar, Tony Gaze flew to Eindhoven to seek out everyone he knew who could help him with a posting to an active squadron. His operational hours were now 531, and claims were 9.5-4-4. After some time at 83 Group Support Unit, he was posted as commander of A Flight 41 Squadron, arriving on 6 April. The squadron was flying the Spitfire XIV-E, much improved and with the gyro gun-sight which greatly aided air-to-air shooting, as well as .5-inch machineguns in place of the .303s. Few Luftwaffe aircraft were met, but, for Gaze, the final six weeks of the war were the most intense operational flying of his career, as 41 Squadron concentrated on the Reich railway system. Its Spitfire XIVs swept out to Nienburg, Bremen, Hamburg, Hoya, Lubeck and even to north of Berlin. Two or three sorties were flown on the 7th, 8th, 9th, 10th, 11th, 12th, 13th, 14th, 16th, 17th, 18th, one on the 19th, 20th, 21st, 22nd, one on the 23rd, four on the 24th, three on the 25th, three on the 26th, one on the 28th and 29th and three on the 30th. Trains and trucks were left burning and smoking behind them and nothing was safe. Sometimes there was no flak, sometimes plenty. He flew some 80 operational hours in three weeks. On 10 April, Tony saw an Ar234 and gave chase, but the jet sped away. Then he saw five FW190s but could not get to them, and had to break away from another nine as he had no petrol for a combat.

On the 11th, after strafing three trains, he attacked and

shot down a Junkers Ju52 three-engined transport—the German equivalent of the faithful Allied DC3; burning, it crash-landed. He flew EB-E back to base and returned to the same area, Bremen-Nienburg, in EB-R. Another Ar234 was seen, chased and fired on at 800 yards (metres). There was no visible result, but Tony believed he hit it. The German flew on, out of sight.

The Ar234s were active in the area and next day, after a morning sortie in which a train was strafed, Gaze returned to the Bremen-Hoya area in EB-E and, with Flight Lieutenant D.V. Rake, chased another Arado, forcing it to crash. German flak did its best to interfere, but was unable to hit the Spitfires.

'I shouldn't have got the Arado, really,' recalled Tony, 'because the chap didn't know what he was doing and tried to outclimb us. If he'd stuck his nose down, he'd have gone. The reason I shared it was, it was almost impossible to shoot at,

Arado AR234, a German jet type shot down by Tony Gaze. (RAAF official)

with the gyro gun-sight. It got you absolutely plumb astern on a chase, but you got into the slipstream or vortices off the wingtips, and you could only get a ½-second burst, then you were twisted about. With an ordinary sight, you wouldn't have been that accurate, exactly behind. I kept doing little tiny squirts, knocked one engine out, and the other was smoking, so I don't think he'd have got very far. I certainly would have finished it off, but ran out of ammo. He stalled when I was shooting at him; my No.2 followed him down and as the Arado tried to make a crash-landing, he blew him up.'

Still the German people suffered the whirlwind sown by Hermann Goering in 1939. The Allied bomber formations continued to strike at targets in the areas not yet reached by their armies. Kiel was attacked on 3 April, leaving 414 killed, 119 wounded and 12,000 homeless; on the 9th, leaving 82 dead, 73 wounded and 20,000 homeless, and on the 13th, with 90 dead, 41 wounded and 2,700 homeless. Other towns and cities were pounded, while fighter-bombers attacked anything on the roads, railways and canals.[9]

On 16 April, 41 Squadron moved to the Luftwaffe base at Celle. On the 17th Tony attacked a Ju88, through flak, but ran out of ammunition, and Flying Officer Hegarty destroyed it. On the 19th Tony found himself chasing a jet, later identified as a Me262; he looked around and saw what was thought to be a V-1 flying as if in formation with the 262. Afterwards, it was decided the 'V-1' was a Heinkel He162 'Volksjager'. The jet pulled away.

Next day, German aircraft were reported, but he had to leave the area without enough petrol to go looking for them. On another sortie, he was hit in the radiator. Next day, the 21st, flying through German flak of 88, 37 and 20mm, he was hit again in the radiator. At these times, the pilot owed his life to an alert No. 2, who would call a warning about trailing glycol; at that point there was nothing to do but turn for home and go flat out, to at least get out of the area of the people who had just been on the receiving end of the strafing attacks, and hope the engine life extended to the flight time necessary to return to base. Once glycol began to pour out it was useless to try to nurse the engine. The radiator was hit again on the 24th, during a morning sweep. In the afternoon, in EB-F, the Spitfires met Soviet aircraft. The Russians did not hesitate, but went for anything not their own and Tony 'nearly clobbered a Pe2!' Other pilots did fire when fired upon. Next day, another hit in the radiator.

More Luftwaffe aircraft were met on the 28th, when four Spitfires attacked ten FW190s seen circling Schwerin. The 190s jettisoned the bombs they were carrying and Tony shared one which spun into the ground; then, as he was about to fire on a second, Flight Lieutenant Wilkinson 'cut me out, damn it!', destroying the Focke Wulf. On 30 April, during a late afternoon patrol over the Allied bridgehead on the eastern bank of the Elbe, there was a combat with FW190s and Bf109s. Squadron Leader Shephard claimed one of each type, and Tony claimed a FW190. During the day, 41 Squadron destroyed 19 aircraft.

The warrior's sense of humour during this time cost a friendship between two Canadian pilots. In the midst of a fierce tussle with German fighters, one of the Canadians found himself in trouble, called for help, and then said, 'They're all around me! They're all around me!' The other Canadian called with, 'Are they pickin' on ya, Stanley?' The pilot was rescued, but Tony recalls that they were no longer friends.

At the beginning of May, Gaze was again posted to 616 Squadron. It was equipped with Gloster Meteor jets, the first

Allied jet to see operational service. A flight commander, Squadron Leader L.W. Watts DFC, had been killed in a collision, and a replacement was required. Tony wrote in his log book, 'Farewell to Spits, bugger it. Squirts of all things!'. He took command of A Flight, at Fassberg. After testing one, he wrote, 'quite a nice kite. Bit heavy though.' After one 30 minute flight, he next flew the Meteor on operations. The squadron had to move continuously, as the range of the Meteor III was only some 45 minutes, 'and not 46 minutes', as he emphasised. 'This range business is no joke.'

The squadron moved to the former Luftwaffe bases at Luneberg, then Lubeck, and the RAF members were surprised to find the German catering and messing staff, including servants, stayed on and continued in their old jobs, now for people wearing a different shade of blue uniform. They seemed to be quite happy to do so.

Tony flew six operational sorties in Meteors, destroying or damaging some 15 German vehicles.

The last was on 4 May, and then the war ended.

He had flown 488 operational sorties, claimed 12.5-4-4 enemy aircraft, and had 1636 hours in his log book.

With the shooting stopped almost everywhere, but the war not yet officially over, he took off in a Meteor, flew to the stretch of autobahn in Shleswig where Messerschmitt 262 jets were gathered, and landed. Officially this was still a Luftwaffe airfield, as the British Army units were some miles away, and he knew he was breaking all sorts of rules and regulations. However, the Germans were not hostile, and the commanding officer appeared. Speaking with the authority of the victor, Tony ordered a 262 prepared for him to fly. The talk turned to the other types he had flown and conversation moved to wartime aviation. This unit had the Me262 night-fighter, with a second seat and radar added. The commander was probably Oberleutnant Kurt Welter, Knight's Cross with Oak Leaves, believed to have scored about 50 victories, many in command of the unit, 10/NJG11. Welter—if it was him—was born in 1916, and would have been 29 years old. Tony thought that he seemed quite old, and all realised that the Germans would not be able to fly anything for some considerable time.

After instructing Tony in starting and handling procedures for the Me262, the German asked if he could sit in the cockpit of the Meteor. Tony realised that if he was in trouble up to that moment, he could go to jail for allowing an enemy officer into the cockpit of the latest and very secret product of British aviation. But he took the risk, and the German was quite impressed. The 262s at that stage 'were thrown together, the instruments did not match', as Tony noticed. The 262 was started by a cumbersome procedure of auxiliary generators, a small two-stroke engine in the 262 nacelle being brought to life to start turning the Junkers Jumo turbine. By contrast, Tony started the Meteor himself while talking to the German pilot. Another feature which brought questions from the Germans was the lack of a gun-sight poking up and out from the instrument panel. Tony used the right switches and the sight pattern appeared on the windscreen, again bringing mutters of approval from the onlookers.

Welter warned Tony about the dangers of flying the two-seater Messerschmitt, saying that they were deadly. The Messerschmitts had been chasing high-flying Mosquitos, with some success. They thought the Mosquito at night was relatively easy prey, as the RAF crews assumed their speed was protection enough and they did not seem to look behind. It was also obvious to Tony that the Germans 'had Mosquito intruders on the brain, and were really scared taking off and landing that a Mossie would sneak in behind and get them.'

Then Welter explained that they knew the British Army

would be there next day, and they would be prisoners, but it was intended to have the biggest and best party possible, with all the unit 'and the Lustwaffe', and Tony was invited to stay as guest of honour. He was mightily tempted, but thought that 'later in the night some rotten little Hitler Youth will stick a knife in me, so I'd better not. But it would have been good fun.' He took off and flew back to Lubeck.

As the war ended, and for a few days after, Luftwaffe aircraft would arrive, sometimes with groundcrew aboard, flown by pilots quite ready to continue the war against the Soviets, and who were a little surprised to be informed that their operational days were over.

Most people felt an enormous relief that the war in Europe was over but others, Tony included, started thinking about the war in the Pacific. It seemed far from over, and the RAF had begun establishing a series of bases to enable aircraft to be flown across Europe, the Mediterranean, India and Asia to take part in the final campaigns against Japan. The Royal Navy came around, hunting for pilots with twin-engine experience to fly the Sea Hornet.

Meanwhile, in the strangely peaceful skies over a sunny Germany, the British and Americans—and Russians—were enjoying the experience of flying many of the machines formerly used against them. Focke Wulfs, Messerschmitts, Junkers, Heinkels, Buckers—anything which could fly—was painted with the markings of new owners and then tried out.

German pride in their work extended to carefully painting RAF markings on 616 Squadron's FW190. As Tony commented, if the situation had been reversed and the Luftwaffe required black crosses painted on Spitfires by RAF groundcrews, the result would have been a crude horizontal and vertical sploshy brush stroke. On 28 May he flew a Messerschmitt Bf108 to collect a Siebel 204 twin-engined aircraft.

Tony Gaze in Spitfire cockpit. (Tony Gaze)

The ground staff there, all German, refused to allow him to take it until it had been checked and brought to an acceptable standard, including having the perspex polished. Then his confident claim to knowing the cockpit drill was ignored and he was instructed in starting and take-off procedures— finding out that his previous self-taught drill was in error and he had been flying another 204 with both engines drawing from a single tank. All this was done in spite of his feeling poorly from the night before, his breath filling the cockpit with boozy fumes, drawing a smiling 'schnapps?' from the tutors. When the Germans were relatively confident that he would be able to remove the aircraft safely from the area and presumably to a distance where they could disclaim any responsibility for any crash, he took off, did a mild beat-up and left a friendly waving group behind.

The Siebel was then taken and used by 616 instead of the more normal Avro Anson, and its fine flying qualities and design features were appreciated by all. The first Seibel they had acquired had been found in a field, with flying controls disconnected—as required by the surrender terms. After replacing the propellors and re-connecting the controls, the engines were started and Tony took off, out of the field, to fly at low level to Copenhagen. All along the way, he could see people shaking their fists at him as he passed, and he wondered why this should be. It was only after landing he realised black crosses were still on the wings and fuselage.

In the free spirit of the times, Tony flew a used FW190 to a place where there were brand-new ones collected, and persuaded the guards there to let him swap. After all, they still had the same number of aircraft on charge. His greatest regret is that he never actually flew a Messerschmitt 262. Two were collected from Fassberg but he did not fly either, believing that there would be plenty of time to do so. The squadron commander had taken off in one, and Tony was preparing to go in the second, as the first came back for a landing. Kurt Welter had impressed on Tony the long time needed for the nose wheel to come down, a feature of the 262 which required a long landing approach. But the squadron commander did not allow for this, and the wheel was not fully down when the weight of the aircraft came onto it, with the result that the nose went down, and both engine nacelles scraped along the runway, rupturing the small fuel tanks for the two-stroke starter motors housed there, starting fires. Tony was taxying, and waited for the landing before he took off himself, but was ordered not to go ahead. 'It was infuriating. If he'd been two minutes later, I'd have been in the air,' he said.

But soon the use of German aircraft was officially banned, as most of them were flown without maintenance and presumably there were crashes from that and other causes. In addition, Allied technical and design teams were scouring the country looking for examples of German design and many aircraft were taken back to the UK, to the USA and on the other side of the Russian line everything moveable was shipped east.

At the time, little was known about the effects of compressibility and the sound barrier, though some of the faster aircraft had encountered problems, particularly in dives. Up to the end of the war, the Meteor pilots had been ordered not to use the dive brakes, and, as Tony recalled, 'were bending the aeroplanes all the time. We ended up in our own little cloud formed by the shock wave. Then, as soon as the war ended, you could use the dive brakes.'

Gaze decided to investigate the sound barrier. 'I decided to find out what the speed of sound was like. I thought, "I can't see any reason why you can't dive through this Mach number". I climbed as high as it would go, rolled it onto its back, and aimed down. I soon found out. I don't think I went down more than a thousand feet, and everything flapped and banged, and it wasn't going to go through it. The blooming wings nearly fell off. I chopped the throttle and pulled out at 25,000 feet and looked behind. There were two great vapour trails, which I hadn't made going up. They were unburnt kero, and I'd blown the fires out. Luckily, I was aiming back to Lubeck. I didn't know what was the best glide angle for a Meteor, so just aimed the nose around until Lubeck started going down, not up, and I knew I'd get there. When I came in, the cockpit iced up completely as I got into the circuit, but luckily I'd flown there a bit, and wound the hood back, and came in looking sideways. I hadn't read the book, and the book said, "Do not try to relight above 10,000 feet." I'd been trying way up high.'

Tony wrote a report on the episode, but wondered if it had

been circulated, as he later heard of an RAAF pilot whose aircraft was lost in similar circumstances.

Tony was used for a time to teach people to fly jets, then sent to Rolls-Royce, who wanted a three-year commitment which he would not give. He wanted to go back to Australia, and for a time it seemed that he would be able to do so by flying a Spitfire XIX in a record-breaking attempt along the chain of bases established by the RAF for continuing the war against Japan. Then at the last minute, that was cancelled. He was actually packed for the flight to Australia to become the personal aide to the RAF senior officer there, when told the posting was cancelled, as an Australian could not be seen to represent the RAF in Australia. The logic escaped him.

For a time he commanded 691 Squadron, which operated a variety of aircraft, and co-operated with Royal Navy ships in calibrating radars and practising gunnery, then took over a Test Flight, but finally resigned and came home to Australia. He flew with 21 (City of Melbourne) Squadron for a time, but returned to Europe when car racing developed. Without RAAF support, his intention of joining 600 Squadron in the Auxiliary Air Force was not possible, and he resigned for the last time in 1951.

Gaze became interested and involved in gliding, representing Australia in World Competitions in the '60s. On one occasion, he had to go to Germany to have the cockpit of his glider altered to accommodate his tall frame. As he used a British parachute, he had to take this with him so the cockpit alterations would be correct. Tony carried the parachute into the cabin of the British Empire Airways airliner and resisted polite attempts by the hostess to take it from him. Other passengers became worried at the sight of this man who insisted on holding a parachute. The pilot came, reassuring the passenger that a parachute was not necessary, and Tony replied that he'd been flying for years, and 'never go anywhere without my parachute.' The same scene was replayed on his return flight!

At time of writing, Tony Gaze lives on a property in Victoria.

Appendix

Aircraft flown by Tony Gaze most often when on a squadron, or when making victory claims.

610 Squadron

21 June 1941	Spitfire IIB	DW-K	one Bf109 damaged
26 June 1942	Spitfire IIB	DW-H	one Bf109 destroyed one Bf109 probable
2 July 1941	Spitfire IIB	DW-G	one Bf109 damamged
6 July 1941	Spitfire IIB	DW-A	.5 Bf109 destroyed
10 July 1941	Spitfire IIB	DW-G	two Bf109 destroyed (one may be Uffz E. Hammon 2/JG26)
17 July 1941	Spitfire II	P8749	one Bf109 probable (may be Uffz G. Oemler 9/JG26)

By the end of his tour with 610 Squadron, Tony had flown Spitfire IIB DW-G 46 times, and Spitfire VB DW-G 71 times.

616 Squadron

| 13 July 1942 | Spitfire VI | YQ-6 | one FW190 probable (may be Ofw Helmut Ufer 4/JG26) |

Tony Gaze DFC

18 July 1942	Spitfire VI	YQ-6	one FW190 destroyed one FW190 damaged
30 July 1942	Spitfire VI	YQ-G	one Bf109 frightened
19 August 1942	Spitfire VI	YQ-G	one Do217 destroyed

YQ-G flown 96 times.

64 Squadron

6 September 1942	Spitfire IX	SH-G	two FW190 damaged
27 September 1942	Spitfire IX	SH-G	damaged on landing
11 October 1942	Spitfire IX	SH-G	one FW190 damaged

SH-G flown 36 times.

616 Squadron

4 December 1942	Spitfire VI	YQ-S	100th offensive sortie
12 December 1942	Spitfire VI	YQ-S	one FW190 damaged

YQ-S flown 61 times.

Station Headquarters Hornchurch

16 August 1943	Spitfire IXB	DV-A	one FW190 damaged
17 August 1943	Spitfire IXB	DV-A	one FW190 destroyed
19 August 1943	Spitfire IXB	DV-?	one Bf109 probable (may be Ltn Johanes Meyer 10/JG26)

Various aircraft flown throughout this time.

66 Squadron

4 September 1943　Spitfire VB　LZ-A　　one FW190 destroyed; shot down, claimed by Gerhard Vogt JG26; escaped via Spain. Another FW190 crashed in the low-level combat; (both may be Ltn Ernst Heineman 4/JG26 and Ofw Walter Grunlinger Stab/JG26)

Various aircraft flown before being shot down.

610 Squardon

Date	Aircraft	Code	Notes
5 August 1944	Spitfire XIV	DW-U	one VI destroyed
1 January 1945	Spitfire XIV	DW-T	one FW190 destroyed (probably Uffz Werner Hilbert, 9/JG2)
16 January 1945	Spitfire XIV	DW-T	starboard wing hit by 88 mm flak
14 February 1945	Spitfire XIV	DW-F	one Me262 destroyed (Fw Rudolf Hoffmann, 9K+NL, I/KG51)

DW-T flown 67 times.

41 Squadron

Date	Aircraft	Code	Notes
11 April 1945	Spitfire XIV	EB-E	one Ju52 destroyed
12 April 1945	Spitfire XIV	EB-E	.5 Ar234 destroyed (probably KG76)
28 April 1945	Spitfire XIV	EB-B	.5 FW190 destroyed

Tony Gaze DFC

30 April 1945   Spitfire XIV EB-E   one FW190 destroyed

EB-E flown 19 times.

616 Squadron

-- May 1945   Meteor III   ----   flight to Me262 base on the autobahn

Total operational sorties were 488.

The thin red line

Burma Hurricane

Sqn Ldr W.J. 'Jack' Storey DFC MID

SIX ACES

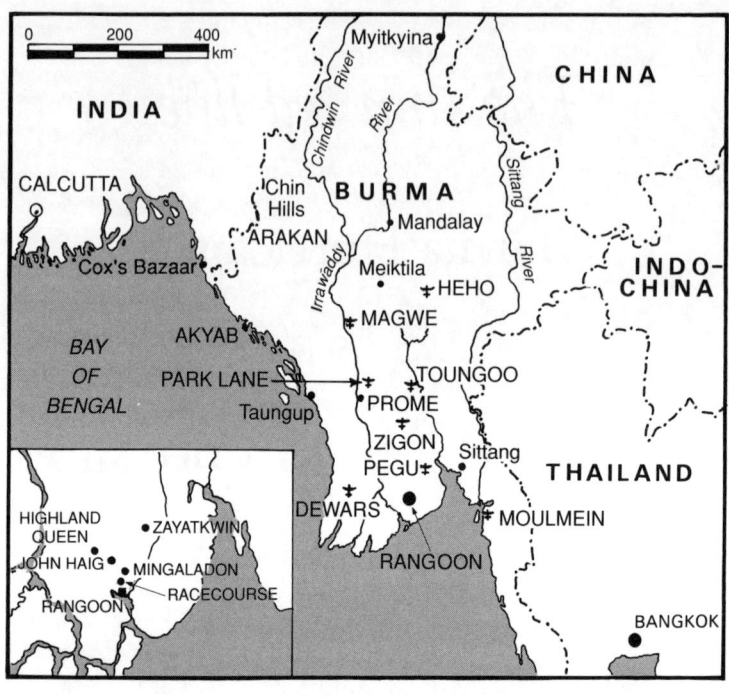

W.J. 'Jack' Storey DFC MID

Before the war, W.J. 'Jack' Storey began his career as a schoolteacher in 1935. In 1941 he began pilot trainng, and later went to Canada and the UK. Eventually, he became a member of 135 Squadron RAF, and late that year the squadron was posted overseas, the intended destination being Russia. However, this plan was abandoned when H.M.S. *Ark Royal* was sunk in the Mediterranean, and 135 Squadron arrived in Cairo in December 1941. After Japan attacked Pearl Harbour, 135, 136 and 17 Squadrons were posted direct to Rangoon.

Squadron Leader Frank Carey commanded 135 Squadron, and he decided that six pilots would go by BOAC aircraft to Rangoon as an advance party, leaving the rest to fly the squadron's fighters along the remainder of the ferry route from Takoradi, on the west coast of Africa, with ground-crews to follow by ship. However, on arrival at Karachi, four of the advance party were removed from the BOAC aircraft, their places being taken by military personnel of higher priority. This left only Carey and Jack Storey to go on to Rangoon, where they arrived on 19 January, in the middle of a Japanese air raid. They noticed Tomahawks of the American Volunteer Group (AVG), the 'Flying Tigers', operating from the airfield at Mingaladon, along with a few Hurricanes, and met the commanding officers of 17 and 136 Squadrons.

These experienced pilots had already flown against the Japanese, and had returned with bullet holes in their aircraft. There was what Jack Storey described as 'great consternation as to how to deal with these Japanese fighters, which were quite extraordinarily manoeuvrable. The conditions of combat were so different from the Battle of Britain that new tactics had to be evolved.'

Air Vice Marshal Stevenson, RAF commander in Burma, had available only one AVG squadron with 21 Curtiss 'Tomahawk' fighters, and 67 Squadron RAF, with sixteen Brewster 'Buffalo' fighters; reconnaissance was to be flown by 4 Flight Indian Air Force, with Westland Wapiti and Hawker Audax biplanes; liaison flights were to be performed by a flight of the Burma Air Force equipped with a variety of Moth aircraft. This small force was assisted by one obsolete radar in the Rangoon area, complemented by hilltop observation posts linked to the control centre by telephone. The network of observation posts gave 20 minutes warning of Japanese approaching across the Gulf of Martaban, but the primitive radar consistently failed to give accurate information of the height of the approaching enemy, often reporting them to be 5,000 feet lower than they were.

Carey and Storey went to Zayatkwin, 30 miles (50km) north of Rangoon, to prepare the airfield for the arrival of the squadron, but the Japanese bombed and strafed it continually, so they returned to Rangoon on 29 January, and went on readiness in two Hurricanes which had arrived at Mingaladon.

An estimated 5,000 Burmese had been killed in a series of daylight raids made on Rangoon by the Japanese between 23 and 25 January. On 28 January, the pilot of a damaged Japanese fighter had realised that he was not going to be able to return to his base, so tried to deliberately crash into a parked AVG Tomahawk. He missed, and went into the ground next to the pen. Later in the war, the Kamikaze attacks received great publicity, but from the first days of hostilities, Japanese pilots and crews who found themselves unable to return to base were encouraged to ram enemy aircraft or crash-dive into ships or other targets, in what was called 'self-destruction'. Landing and delivering a damaged aircraft to the enemy, or surrender, were an anathema.

On 29 January there was 9/10th cloud cover, through which Carey and Storey were scrambled against a reported six enemy aircraft flying at 12,000 feet. A veteran of the

Battle of Britain, Carey knew the value of height, and they climbed to 17,000 feet. At 12,000 feet he saw the Japanese Army Ki27s against the clouds and led Storey down at them.

Rudder marking from Japanese Army fighter of 77th Sentai, shot down January 1942. (Jack Storey)

The first attack failed, as the Japanese flew into more cloud, so the Hurricanes used their speed to climb and then rolled back down. The Japanese came out of cloud and began to circle; two of them lagged and became separated from the rest. Then the AVG Tomahawks arrived and began dog-fighting with the Japanese.

From his diving Hurricane, Storey saw three Japanese 'Nates'—their Allied recognition name—queued behind one Tomahawk, easily out-turning the American. Carey came down in a steep right hand spiral, at 310mph (500kmph) onto the Japanese. Storey fired two bursts and hit a Ki27 Nate, which swerved left into cloud; he followed. The groundcrews could hear the firing. As Storey came out the other side of the cloud, diving over the base at Mingaladon, the Burmese-manned anti-aircraft defences opened up. Storey quickly broke away, last seeing the Nate low over the airfield, and five minutes later he came back to land. The Japanese had crashed into a blast pen, just missing a Blenheim, evidently trying to die striking a final blow for the Emperor.

Two Tomahawks were lost, and a third force-landed with a pilot wounded in the head and leg. In the wreckage of the Nate, it was found that the Japanese pilot was shot, wounded in the back, by either Carey or Storey. Neither pilot claimed the Nate, but it was later awarded to Storey. He believes that this was a gesture of generosity by Carey, who already had some 18 victories, and was to end the war with about 28. Among the possessions on the dead Japanese pilot was a sketch map of the airfield, with the position of all parked aircraft marked by crosses. It was assumed that this sketch had been made by referring to photographs taken by high-flying Japanese reconnaissance aircraft.

Scrambles and combats followed throughout the following days, with sometimes as many as five sorties a day. The AVG were as heavily involved as the RAF, and each preferred their own type of aircraft—Tomahawk for the AVG and Hurricane for the RAF. However, the numerical superiority of the Japanese meant that their losses were not as significant as for the defending fighters, and no matter how many were shot down, it was plain that the Japanese were winning the campaign for Burma. On 1 February, the airfield was bombed for some five hours by Japanese aircraft.

Barry Sutton, who later commanded 135 Squadron RAF described, in his excellent *Jungle Pilot*, how, before engaging the Japanese in action, he had read a newspaper article by an authority on the Asian peoples. The writer had explained that as Asian skulls were of a particular shape, the centrifugal force of any aerial manoeuvre pressed the brain against the skull and the Asian promptly lost consciousness; they could not be considered a menace in the air. Reality was far different, and Sutton often wished he could have had the said 'authority' in the air, combating Japanese fighters which made extremely tight turns and other moves.

It was noticeable to Storey, and other relatively inexperienced pilots, that many senior pilots and leaders, who had flown in the Battle of Britain and sweeps over France, were shot down very quickly, sometimes a few days after arrival. A similar phenomenon had been noticed in the squadrons battling to defend Malta. Observing the Japanese, it seemed to Jack Storey that, 'in spite of their inferior armament, they were exceptionally well-trained in deflection shooting. Our aircraft would come back with holes all over them sometimes. That's difficult to do.' Of course, these Japanese had the benefits of lengthy peace-time training, plus the invaluable experience of combat against the Chinese since 1931 and against the Soviets in the Nomohan Incident in 1939. 'They were good,' recalled Storey, 'or they'd never have got these results. At Rangoon in January and February 1942, they were outstanding.'

Tropicalised Hurricane flown by Jack Storey, Burma 1942-43; WK·C was flown while gaining all his victories. (Jack Storey)

Apart from the modern airfield at Mingaladon, conditions at other bases were quite primitive. The heat was constantly high. Aircraft parked in the sun soon had metal parts too hot to touch, and on occasion the de Wilde .303-inch incendiary machinegun ammunition in the wing magazines of the Hurricanes detonated from the intense heat. Coconut matting had to be placed over the the wing magazines, and continually dampened with water. If bare arms touched the cockpit canopy runners, along which the canopy slid, the heated metal would raise blisters on the arms of the unwary. After the refuelling tankers had been destroyed by bombing, refuelling had to be done from tins, with the petrol filtered through a chamois. Tools and machine shop facilities were almost totally lacking, but the Burma Railways' staff gave free use of their facilities to the RAF units, even assisting where possible.

As well as carrying out interceptions, the Hurricanes escorted RAF bombers and strafed Japanese positions around Moulmein. Only Frank Carey and Jack Storey were trained as night fighters, and some of the other pilots—last seen in Cairo—had been diverted en route, going to places such as Colombo. The result was that, including those lost on operations, 135 Squadron was reduced to six to eight pilots who were flying every day. Night bombing disrupted their sleep and rest so that tiredness and, finally, exhaustion, became a major problem.

The incessant Japanese attacks finally exploded the armoury at Mingaladon. After Moulmein was lost, the air warning net was broken and the threat to Rangoon from air attack increased greatly. The aircraft were flown away from the obvious target of the major base at Mingaladon and dispersed to a series of strips carved from the dry-season paddies. These were called 'kutcha' strips, and were code-named, even officially, after popular brands of scotch: Johnnie Walker, Highland Queen, John Haig and Dewars. Naturally, the kutcha strips were very dusty, and the hard sun-dried earth cracked, producing a lumpy but level runway. However, the cracks in the soil became so large that the Hurricane tail wheel could—and did—drop into them, sometimes tearing off and causing damage to the tail area. The obvious field modification of fitting locally produced tail skids in place of the wheels was tried, with some success.

The pilots were quartered some 25 or 30 miles (48km) away and would rise at about 04.00, drive to the kutcha strip, fly to Mingaladon for whatever operations were required during the day, then fly back at dusk to the dusty strip, and perhaps be required for night operations. If rostered for night operations, being on readiness involved being on duty all night until dawn. On one occasion, Storey was on duty for 24 hours continuously.

On 6 February, Storey led five Hurricanes against 30 Japanese fighters, engaging them just north of the base. Two of the Hurricane formation became separated during the climb to engage but, when at 16,000 feet with the other three, Jack Storey saw the Japanese above them, at 21,000. He led the small formation to the right, into the sun, still climbing. Then, at 18,000 feet, they were seen and attacked. Three Japanese dived on them from 1 o'clock—ahead and slightly right. (The Allied air forces used 'the clock code' to indicate relative positions, with the aircraft as the centre of an imaginary clock, so that 12 o'clock was straight ahead, 3 o'clock to the right, 6 o'clock directly behind, and so on.)

Climbing steeply left, in a spiral into the sun, Storey turned and fired at one Japanese, then a free-for-all developed. He got onto the tail of a Nate and fired from dead astern; it caught fire and went down vertically, crashing south-east of Zayatkwin. Storey repeated the steep left climbing spiral tactic and engaged other Japanese, then put two bursts into another Nate, which spun to the right, crashing east of Zayatkwin. He made other attacks, and had a third Nate in his sights, but was out of ammunition.

Storey returned to base with the Hurricane damaged by Japanese fire—a bullet in the starboard mainplane. To ensure top engine performance, he had all the spark-plugs changed as well. It had been a successful combat, with Storey claiming two destroyed (subsequently confirmed) and two damaged. Pilot Officer Underwood RAF also destroyed one Japanese. All the RAF fighters landed out of ammunition. Sergeant Malcolm McRae RAAF destroyed a Japanese aircraft in the combat, but was wounded by fragments of his armour plate which were flung off the far side of the armour by the impact of Japanese ammunition. This was something which the pilots who had been in action against the Germans and Italians had not seen. Though the Japanese fighters only had two machineguns, their ammunition seemed to have more penetrative power.

Storey found that above 13,000 feet, the Hurricane supercharger gave the RAF fighter an advantage in speed and climb, but lower down the Japanese aircraft were superior. The Hurricane II was fitted with the Rolls Royce Merlin Mk.XX engine, which had a two-stage supercharger. Above 13,000 feet, the supercharger was spun at higher revolutions, to help compensate for the decreasing atmospheric pressure. The control was located in the cockpit, for use by the pilot. At low altitude and low speed, the Japanese fighters were very manoeuvrable, and could out-turn a Hurricane. 'The obvious thing not to do,' Jack Storey recalled, 'was to be lured into a low level dog fight at low speed. Ideally, one would want to fight at high altitude and high speed, where the Hurricane was more in its element. We couldn't always get those conditions.'

Storey realised that the obvious course was to make the best use of the flying characteristics of the Hurricane. Confident in the strength of the Hurricane, and aware that the lighter construction of the Japanese fighters probably did not allow them to match that strength, he flew accordingly. If a Japanese was on his tail, he found that 'the best evasive action was to immediately roll and dive vertically in a spiral, to make his deflection shooting difficult as the speed built up, keep spiralling, keep vertical until the speed went up to something over 400, 415, 420, then immediately pull hard back on the stick, put as much G on the aircraft as your body could stand. You didn't black-out but went right up to grey-out and rocketted upwards vertically. If a Jap had followed you in these conditions, you knew that when you straightened out to rocket up again, he could not take a deflection shot on you, because in order to do so he would have to pull even tighter G to bring his sights ahead of your aircraft to lay off

Jack Storey, Hurricane pilot, shot down eight Japanese over Burma, 1942-43.

the deflection. This meant he would either pull so hard that he blacked himself out and couldn't see you, or he'd greyed out and could not pull enough deflection to hit you. I used these tactics over and over again, and never once for the rest of the war was I ever hit in a dog-fight by fire from an enemy aircraft.'

On 5 and 6 February, the Tomahawks and Hurricanes destroyed ten Japanese aircraft, and probably destroyed ten more, for no loss. Two more were shot down at night. On the night of the 7th, Storey scrambled twice against Japanese bombers, but poor fighter control did not bring him into contact with the bombers, who hit the petrol dump at Zayatkwin, as well as damaged the runway and flarepath. The night flying was in addition to the normal daylight operations, and on occasion Carey and Storey scrambled twice at night. Once there were no lights to land by, and they were forced to do so by moonlight, trusting to luck to avoid bomb craters. The AVG made an attempt to fly at night, but abandoned the idea after the first aircraft crashed on landing.

The weeks went on, providing escorts for Lysanders and Blenheims and carrying out interceptions by day and night, amidst constant Japanese attacks. All the while, the squadrons were operating in a hot, humid climate, with little rest. One aspect of the campaign which caused a great deal of thought to Jack Storey, from a Dominion of the British Commonwealth, was the smallness of 'the thin red line' of defenders in Burma. 'It was damn thin,' he said. 'It was not a fighter force at all. You can't carry on a war with that (small a force), when you're being invaded by scores and scores of highly organised and skilful, experienced Japanese pilots. The impression one got out there was of operating in isolation. We'd left Europe, with comfortable Messes, sufficient food, (and) reasonable climate, to go to a place full of malaria, dengue fever, little food, hard-to-get spares, shortage of pilots, and the Japanese coming, literally, in swarms. It occurred to one that this is the RAF Far East Command; this is what's trying to hold the line; to kick the Japanese out of India. It just seemed so pitiful.'

The second city in Burma — Moulmein — was attacked by Japanese aircraft on 18 February. Three days later, a large conglomeration of 300 Japanese trucks, cars and carts was seen on the Bilin-Kyaikto road, and it was the Japanese turn to be on the receiving end. 38 RAF fighters, with eight Blenheim bombers, attacked the enemy concentration. The Japanese were checked, but not halted, and went on to capture the Sittang River line, seriously threatening the British ability to hold Burma.

It became obvious that, if Rangoon fell, air support for the campaign could be organised and provided only from India. A base organisation could be established there, with forward fields at Akyab, Magwe and some others closer to Rangoon.

On 23 February, Storey was in a flight of six Hurricanes over Kyeikto. Frank Carey saw a Japanese reconnaissance aircraft low over the trees, presumably looking for British ground forces, so he dived to attack. Then, on the left, Jack Storey saw six Japanese fighters approaching, and swung into them. One Nate was lagging, out to the right, turning slowly away from the others, and Storey thought he may have been looking down for the reconnaissance aircraft. He flew up behind the Japanese and fired. The Nate burst into flames and went down burning into the trees; he probably never saw the Hurricane. 'He was a pushover,' recalled Storey. The combat took place some 40 miles (85km) east of Kyaikto.

The Japanese air force had tried to achieve air superiority over Rangoon, but the AVG and RAF resisted strongly, providing such a setback that no further attempts were made until Rangoon fell to ground forces. However, on some occasions over Rangoon, the Japanese so outnumbered the defenders that 'we had to look after ourselves', as Storey recalled. The intensity of the operations can be gauged by the log book entries:

> 24 February, scrambles; 25th, scramble; 26th, four scrambles, nearly jumped diving to attack two 97s, aerodrome bombed, dog-fight, got nothing.

The pilots were becoming exhausted. On one occasion, the scramble bell went, but some were slow in getting to their feet. Barry Sutton, now commanding the squadron, said, 'Come on you bastards, do you want to live forever?' Jack Storey recalls that they ran out to the aircraft, revitalised.

The Americans and RAF fought together, sometimes using a wing leader, but the different radio frequencies in the Tomahawks and Hurricanes meant that one of each type radio had to be installed in the operations control room.

Sometimes a Hurricane flew with the Tomahawks, to pass on information.

'Because the Tomahawk had a slower rate of climb,' said Jack Storey, 'we used to allow them to take-off first. We each had our own runway, and the RAF would then take off, climb, and catch the AVG at 12,000 feet, on the way to 25,000. Some pilots wanted the Hurricanes to go first, but it was decided by senior officers to let the AVG do so.' Separate squadron take-offs were necessary since, although each used a different runway, the runways intersected, so could not be used simultaneously. If possible, enough height was gained to allow diving attacks on the Japanese but, if not, the defenders tried to delay engaging until the Japanese were closer to Rangoon, or until the Japanese turned for home, short of fuel. It had been a hard-learned lesson — but by this time accepted — that the nimble Japanese fighters were at their best in a dog-fight and turning match. It had also been learned that while the Japanese bombers had front, top and tail guns, they were poorly protected from below.

Another Australian in the theatre was 'Bush' Cotton, 17 Squadron. On 27 February, Cotton had made several passes at a formation of Japanese bombers, and was preparing for his third dive, having destroyed one, when he noticed some Japanese fighters. As he banked to begin his dive on them, one Japanese pilot pulled up the nose of his Nakajima and fired a burst which hit Cotton's Hurricane around the cockpit, wounding him in the leg. Cotton dived through the Japanese, and managed to land at base. This was another example of the ability of the individual Japanese fighter pilot at this stage in the war.

The warning system for Japanese raids, used to scramble the RAF and AVG in time for them to intercept with an advantage, had broken down, and the Hurricanes and Tomahawks were forced to move away from Mingaladon. A system of standing patrols had begun, with the complementary problems of increased flying hours, fuel consumption, and pilot tiredness. The AVG went first, to Magwe on 28 February, followed by 17 and 136 Squadrons, leaving 135 at Mingaladon with less than half-a-dozen Hurricanes.

Some books written at the time, usually by Americans, claimed that the AVG was left alone at Mingaladon, but the reverse is fact. The Americans left first, and the RAF pilots were greatly surprised by this. At squadron pilot level, there was little, if any, of the tensions which may have existed at command level. Jack Storey recalled that both groups got along well. 'We didn't see much of them, but the ones we did see seemed to be good chaps. They had their own Mess, and supplies we couldn't get. The quality of food, clothing, and so on that they had, compared to ours, well . . . they were not comparable at all. It is a fact that reporters always went along to the AVG and stayed with them, simply because of the living conditions.'

On 28 February, available fighter strength to the defenders was less than ten. Two days before, the situation had been 20 fighters and 16 bombers. Japanese strength was at least 150. However, for those able to consider the situation from more than a personal aspect, there were glimmers of hope. Even with the crushing numerical superiority enjoyed by the Japanese, they did not seem to be aware of the positive results to be gained by turning that air power onto the roads and waterways along which Rangoon was being evacuated. The RAF and AVG presence seemed to be enough to cause the Japanese to act with caution, and the Japanese did not seem to have a correct appreciation of the situation, either tactically or strategically. Even at such a time, when Burma was being given up to the invaders, there was reason to believe that the future would be better.

Next day there was another combat, but the Japanese so

out-numbered the Hurricanes that no claims were made, the RAF pilots twisting and turning in the midst of the enemy, hard put to survive.

On 1 March, Jack Storey was promoted to Acting Flight Lieutenant, as he had been performing the duties of a flight commander since that officer had been killed. Barry Sutton had nominated Storey for the position which involved promotion direct from Pilot Officer to Acting Flight Lieutenant. For Sutton, the choice was an obvious one, and he wanted to acknowledge Storey's exceptional ability as a fighter pilot and one of the best formation leaders he had seen. On that day the administration staffs left Rangoon, from which evacuation had been in progress for some days.

135 Squadron had been operating from nearby kutcha strips: Johnnie Walker, John Haig and Highland Queen. Japanese strafing attacks forced a withdrawal from the area, to Zigon, south of Prome. The Hurricanes mainly attacked Japanese ground troops advancing from Pegu to cut the Rangoon-Magwe road, and though all were hit by return fire, no one was shot down. They attacked targets in the area covered by Moulmein, Kyaikto, Pegu and the Salween River bridge, and on 4 March found 60 Japanese tanks west of Pegu, heading for the Rangoon-Megwe road, which they cut next day. Mandalay was also cut off by road.

By this time, the squadron was reduced to five serviceable aircraft at most, and those were kept flyable by every trick of improvisation and ingenuity of which the groundcrews and Senior NCOs were capable. Eight Hurricanes were available, but some had to be used as spare parts. Eventually, on the suggestion of Sergeant Cummins, in charge of maintenance, it was agreed that the pilots would fly what was provided by the crews, but documentation certifying the aircraft was prepared for flight would not be produced. Sergeant Cummins later received the British Empire Medal (BEM) for what Jack

Storey described as 'his sterling work' in keeping the aircraft flyable, if not officially 'airworthy'. This repair work was done in the open and, on occasion, under fire from snipers creeping around the airfield perimeter.

On 5 March, Japanese barges which had come across the Gulf of Martaban were attacked south of Rangoon, but on the 6th the Hurricanes were caught on the ground at Highland Queen and strafed by six Japanese, who put holes in some of the aircraft although none were destroyed. These holed aircraft had to be burned later when the squadron left the area. The fall of Singapore on 15 February, closing the campaign which cost the British forces some 9,000 killed and 130,000 captured, released more aircraft to the Japanese in Burma. The 30 Allied fighters and 12 bombers were faced by an estimated 400–500 enemy.

Over Rangoon, a towering pall of smoke rose to 15,000 feet and below it a motorized convoy some 40 miles (65km) long crawled away from the city. Japanese road-blocks had been smashed, the force made its escape, and Japanese air units allowed this easy target to proceed with no hindrance. With Rangoon occupied by the Japanese on 7 March, the British and Chinese armies had lost their vital sea supply route. Support for forces in Burma and for the Chinese further north was now reduced to a mere trickle.

By this time, Frank Carey had been promoted Wing Commander and was sent to organise Magwe. The Commanding Officer of 135, Squadron Leader Barry Sutton, was ill. Only Squadron Leader Elsdon, CO 136 Squadron, and Jack Storey, were capable of further operations. In the previous 28 days, these three men had shared the leadership of all formations larger than four aircraft, and each had flown some 60 hours on operations. 'Everyone was sick in some way,' said Storey, 'but, most of all, we were overcome with exhaustion. You were flying such long hours. You can do that for a week, or even a fortnight. Try doing it for three months. God knows how many accidents occurred because of that situation. I weighed six stone twelve pounds (44kg) when I got out of Burma. Also, you did not know if you would be going on a low-level Army support flight, or to 30,000 feet, with sweaty legs and clothes. You could not wear anything other than shorts and a shirt on the ground. When the aircraft came down from 30,000 feet, and landed, pools of water would form under it, condensation from the cold parts of the metal undercarriage.'

However, on 10 March, Storey was ordered out of the combat area to Akyab, for a rest. Next day, 'Burwing' was formed at Magwe, with a total strength of 45 Blenheim bombers, 17 Hurricanes and what strength the AVG could muster for its squadron. On 21 March, nine Blenheims of 45 Squadron and ten Hurricanes attacked Mingaladon, destroying 16 Japanese aircraft on the ground and 11 in the air. However, the Japanese reacted strongly, sending 230 aircraft to retaliate, after which there were only three Tomahawks, six Blenheims and 11 Hurricanes available. But the RAF fought to the end, engaging more strong Japanese attacks on Akyab on 23, 24 and 27 March, claiming four enemy destroyed and three probables, for a loss of six in the air and eight on the ground.

135 Squadron reformed in India, at Dum Dum airfield, outside Calcutta, and on 27 March Storey returned, going on night readiness. The squadron remained there for the rest of the year, apart from detachments to Vizagapatnam for convoy patrols and to Calcutta, which suffered from occasional Japanese raids. Two more RAAF pilots joined the squadron: Sergeants N.L. Crawford and M. de P. Sykes. On 7 January 1943, Sergeant McRae, RAAF, was killed when his aircraft caught fire on landing, and he collapsed inside the cockpit before he could be rescued.

Hurricane IIBs were supplied to 135 Squadron before it left for the Arakan campaign. These were fitted with twelve .303-inch machineguns, which were deemed necessary in the European theatre to destroy German bombers but, for service in Burma, against lightly armoured Japanese, and particularly the agile Japanese fighters, it was decided to remove the four outboard machineguns.

On 23 January 1943, the CO, Squadron Leader Giddings, who had shot down three German aircraft in the Battle of Britain, was shot down in flames and killed. Four days later, the squadron split into its flights, and Storey commanded A Flight at a secret landing strip on the beach at Hove, south of Elephant Point.

Silt had been washed onto the beach, and when the tide was out a firm level stretch of beach, eighty yards (70m) wide was available. This narrowed to 20 yards (20m) when the tide was in. The squadron personnel lived in typical grass huts, with primitive sanitation arrangements, and at night wild elephants caused some loss of sleep. The new CO, Squadron Leader Ian Bayles, was with B Flight, at a location called 'George', an inland strip north of Elephant Point. Bayles, an Australian, was another Battle of Britain veteran, with at least one victory in September 1940.

The main task for the squadron now was escorting Blenheims bombing Akyab, Ramree Island, Taungup Pass, the Japanese fighter base at Meiktila, and Prome, with some strafing attacks on various targets, including Japanese headquarters at Myohaung. In addition, fighter interceptions were carried out.

A new Japanese Army fighter was met on 20 February — the Nakajima Ki43 'Hayabusa' (Peregrine Falcon), called 'Oscar' by the Allies. Jack Storey and A Flight were near Maungdaw, unaware that B Flight was fighting some distance away. In his diary, Storey recorded that 'Fox was shot down in flames, but bailed out before he was burned too much. Gresham was shot down and killed. No Japs brought down. The new Jap '01' fighters seemed to have had a pretty handy performance.'

On 5 March, 135 Squadron flew to Maungdaw Advanced Landing Ground (ALG), and began flying from there in support of the Army. It was a year since the hectic days of the retreat ahead of the unstoppable Japanese.

At 09.15 on 5 March, A Flight was ordered to search for a Japanese patrol reported to be at 15,000 feet over Akyab. The Hurricanes arrived at 21,000 feet, and saw three enemy behind and starboard, closing. When the Hurricanes turned onto them, the Japanese Oscars split up and dived. The fight went down to 12,000 feet, and more Oscars joined in.

Jack Storey, 'after much tail chasing, finally got an opening and carried out a downward quarter attack on a straggler and got in a beautiful deflection shot.' Opening fire at 150 yards (metres), he saw de Wilde strikes sparking on the port wing, engine cowling, fuselage and tail, and several small pieces flew off and away, then Storey passed 20 yards (metres) behind it, through the slipstream, as the Oscar rolled over and went down vertically through cloud, trailing blue-white vapour from the port wing root. Storey pulled away, but one of the other Hurricanes went below cloud and saw the tail of an Oscar poking out of the water, with a cloud of black smoke hanging over it, just where Storey had his combat.

At 12.05, the Flight was ordered off again and, ever mindful of the need for height, Storey went to 25,000 feet over Akyab. They circled for about 20 minutes, gradually letting down to 22,000. Then a Japanese squadron, in vics of three, at different heights, was seen some 1,500 feet below, and the Hurricanes attacked. Storey took Red and White Sections down, leaving Yellow as top cover. He came in behind the third Japanese, an Oscar lagging behind the other two, held his fire until he was again at 150 yards (metres) dead astern, fired as he closed to 20 yards (metres), saw pieces shot off it, then broke upwards, rolling to look back at the Japanese. The Nakajima was diving steeply, and Storey went down after him, but soon the Japanese was diving vertically, at high speed and at 6,000 feet Storey pulled out of the dive, circled and watched the Oscar go straight down into the trees on the northern tip of Baronga Island and explode on impact. The pilot was either dead or the aircraft was out of control. Storey was able to notice the light green wings and darker green on some of the other Oscars.

Climbing as quickly as he could, Storey went back to 15,000 feet, and then saw another Oscar 2,000 feet below, flying straight and level, apparently not keeping a look-out. First looking carefully around to make sure this was not a trap, Storey dived to attack. Again he came in from behind, opened fire at 150 yards (metres), closed to 20, 'got in a very satisfactory burst' and saw pieces, including parts of the perspex cockpit canopy, fly off. Storey pulled up steeply to avoid ramming, the Oscar rolled onto its back and then the pilot baled out. Jack was astonished, as it was the first time he had seen a Japanese do so. Storey circled him for a brief time, and flew off, leaving the Japanese descending slowly over the sea between Akyab and Baronga Island.

In the Mess afterwards, some were of the opinion that Jack Storey should have shot the Japanese under the parachute, but he argued against it, then found an Australian pilot in 136 Squadron had been machinegunned in his parachute the day before.

It had seemed easy to attack all these formations from above and behind, as the Japanese did not appear to be keeping a good look-out. In the combat, Flight Sergeant N. Crawford, RAAF, fired a two second deflection burst at one Oscar, but saw no results and made no claim. Later, he saw another Oscar approaching from ahead, but some 2,500 feet below. The pilot did not appear to see him, so Crawford waited until the Oscar passed underneath and then dived onto its tail, holding fire until he was 200 yards (metres) away, opening fire and giving the Japanese a 5-6 second burst as he came in to almost collision distance. Large pieces flew off and the Oscar lurched into a spin, disappearing into cloud. Crawford claimed a probable.

Jack Storey was awarded an immediate DFC for these combats, but he was also required to write a report comparing the performance of the British and Japanese fighters. Above 20,000 feet, the Hurricane had superior climb, but below that the Oscar was superior, especially in dog-fighting.

'It became evident as time went on' recalled Storey, 'that the Japanese would deliberately leave a decoy section up to 2,000 feet below their main force in order to trap the unwary. This tactic could be countered if one dived on the decoy at very high speed, fired, and pulled sharply upwards, converting speed into height in order to carry out further attacks. To throttle back and stay to dog-fight could quickly prove fatal when the main force swooped. If you did stay to dog-fight, you simply handed the initiative to your opponents.'

Three days later, still operating from the beach, the Hurricanes flew as escort to Blenheims and a Hudson dropping supplies at dusk, and landed back with the aid of the headlights of one Jeep plus two hurricane lamps. On the 11th, two Hurricanes and their pilots were lost in a combat at Rathedaung. The sorties continued day after day.

Petrol was delivered to the beach strip by boat from Chittagong. The drums were simply rolled off into the water and floated in, being retrieved by the airmen. The water was quite shallow, being only waist-deep some 400 yards (metres) from shore.

Food was an unrelenting issue of bully beef and biscuits,

day after day, for weeks. Water was chlorinated. Accommodation was in 'basha' huts built by local people. There were no mosquito nets, and at night the insects 'were really dreadful.' For some reason, many of the airmen were stricken with malaria and dengue fever, while the pilots were relatively unscathed. Everyone worked and flew many more hours than was considered normal. The aircraft were parked and serviced in the open, and the crews tended them day and night, with only the most basic facilities.

Storey went on an instructor's course at the Air Fighter Training Unit, where the CO was his old squadron commander, Wing Commander Frank Carey, DFC** DFM. Storey returned to 135 in April. The Army was retreating from the Mayu Peninsula, and the Hurricanes operated intensively in support of the withdrawal. Scrambles, escorts and strafings continued, with conditions made worse by the onset of the monsoon. On occasion, they flew lone night sorties, fitted with long range tanks, on moonlight nights, across to the Irrawaddy River valley, looking for Japanese barges. Jack Storey recalled these as 'lovely flying, because the air was always so calm and cool, compared to the daytime, and, of course, perfectly safe. That was, safe only until the Japanese had the unfriendly idea of putting steel cables across the valleys and gorges, and we had a couple caught like that.'

On 4 May, the first operation was as escort to Blenheims, but the route crossed that of six Japanese bombers, who jettisoned their bombs when attacked by 135, but the Hurricanes were unable to make any claims. The Blenheims went on, escorted by 79 Squadron. The Japanese top cover was above 135 Squadron, and came down to engage. The Hurricanes 'had to move pretty smart (sic). The Jap fighter pilots encountered certainly knew their stuff. 79 Squadron had two Hurricanes shot up, and one landed with a glycol leak. We made no claims.'

Later in the day, Storey's Hurricane would not start, and

he told Flying Officer Hawkins to lead. He took off late, Ground Control gave him a course, he flew that way, and saw a formation. 'There ahead, and slightly below, was a squadron of aircraft,' he recalled. 'It was cloudy, and we were going in and out of cloud; you'd get a flashing glimpse of them, and then they'd disappear. We broke out of cloud and I caught up with them very quickly. All of a sudden, I saw these things had radial engines. "Bloody Hell! Bloody Mohawks!" I thought.'

Storey had called on the radio that he was approaching the formation from the rear, and received an acknowledgement from the Hurricanes, but the aircraft in front of him obviously were not 135 Squadron. The Japanese formation flashed through another cloud and then became aware of him. One of the leaders curved around in a vertical bank, to check on the new arrival, and Jack 'saw the unmistakeable plan view of an Oscar. I had time to give him a couple of bursts, then had to hop it.'

Storey later deduced that the Japanese were flying relatively slowly to conserve fuel, on a heading for their base at Meiktila. The radar had followed his flight, and approach to the Japanese, but had been unable to do anything, as the real 135 Squadron was too far away to help. Jack did not know if he had hit the Oscar, and did not claim it, though radar had watched it go down and off the screen. Later, an Army patrol found an Oscar in the area of the combat, near Mount Victoria.

Scrambles and interceptions continued, with Storey's Flight operating off the beach, the most forward operating strip at the time. Living conditions were unimproved, with light from candles or hurricane lamps, and the food available was what reached the outer end of the supply line.

When the squadron was taken off operations, Storey had led every sortie except two, as the CO did not fly with A Flight.

Squadron Leader Jack Storey DFC MID, gunnery instructor, Australia, 1943-44. (Jack Storey)

135 then went to Bangalore, and Storey and Flying Officer Fox, RAF, left to form 1572 Gunnery Flight on 4 August, then returned to Australia. Storey was promoted Squadron Leader and commanded the Fighter Wing of the Flying and Central Gunnery School at Cressy until the end of the war. He also re-wrote the syllabus for the Fighter Combat and Gunnery Instructor's Course for the RAAF.

After the war, Jack Storey returned to teaching, becoming a high school principal. Today he lives in Victoria, where his hobby of breeding Connemara ponies has developed into a major family interest.

W.J. 'Jack' Storey DFC MID

Appendix

Aircraft flown by Pilot Officer/Flight Lieutenant W. J. Storey DFC MID RAAF, 135 Squadron Burma 1942–43.

Date	Aircraft	Detail
29 January 1942	Hurricane IIA Z5659 WK*C	one Nakajima Ki-27
5 February 1942	Hurricane IIA Z5659 WK*C	two Nakajima Ki-27 plus two probables; one later confirmed
23 February 1942	Hurricane IIA Z5659 WK*C	one Nakajima Ki-27
5 March 1943	Hurricane IIB AP894 WK*C	three Nakajima Ki-43
3 May 1943	Hurricane IIB AP894 WK*C	one Nakajima Ki-43

Never appear on my CO's parades!

Mediterranean Beaufighter

Flt Lt Mervyn Shipard DFC*

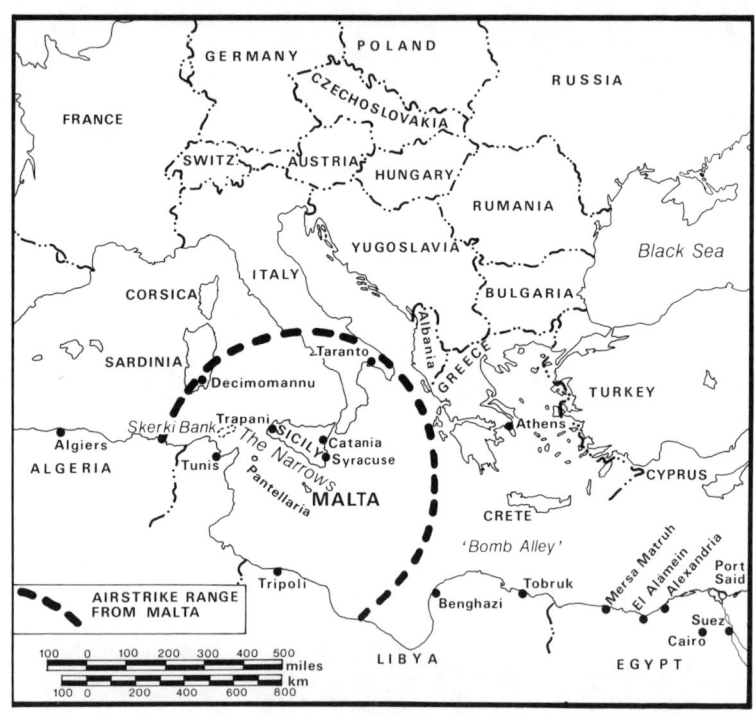

Mervyn Shipard DFC

Mervyn Shipard DFC* RAAF, flew with Douggie Oxby DSO DFC DFM* RAF as one of the most successful Beaufighter nightfighter teams of the Second World War. Douggie Oxby was the most successful Allied radar-operator of the war, taking part in the destruction of 22 enemy aircraft.

Born on 24 July 1917, Mervyn Shipard was the youngest of a family of four. The family owned a property 10 miles (16km) south-west of Henty, between Albury and Wagga. From his earliest days, young Shipard had been fascinated by aircraft, and built several flying models. Carving propellors with the sharpest knife available resulted in a deep wound on each leg, which left noticeable lifelong scars. As well as blood, aviation exacted some sweat, when Mervyn rode his bike from Albury to Melbourne, some 190 miles (310km) to visit the Essendon headquarters of Australian National Airways (ANA). While there, he photographed a brand-new DC3 parked on the tarmac, little dreaming that 10 years later he would be the captain of that same aircraft, taking off from Essendon with a full load of passengers for Sydney.

Young Shipard left school at 17 and worked for the next five years as a clerk with a firm of stock and station agents. He badly wanted to join the RAAF, but his mother resolutely refused to give her permission, so he had to wait until he was 21 years old. He did apply to the RAAF at the end of 1938 but heard nothing until the Second World War began and, on 22 July 1940, just short of his 23rd birthday, he reported to No. 2 Initial Training School at Bradfield Park, Sydney. During the two months there, he developed a close friendship with Frank Richardson, and the two stayed together throughout their training in Australia and Canada.

After flying training on No. 3 Course, Empire Air Training Scheme (EATS), at No. 5 Elementary Flying Training School

(EFTS) at Narromine, the young pilots returned to Bradfield Park and sailed for Canada, arriving at Calgary on Christmas Eve 1940. On the train journey from Vancouver, Mervyn had walked to the engine during a halt, and been invited by the engineer to ride on the cabin footplate during the climb through Kicking Horse Pass, and the spiral tunnel drilled through the mountain.

After settling in at Calgary, and receiving an issue of necessary warm clothing, the Aussies were given Christmas lunch—a magnificent meal of roast turkey with all the extras, served in Service tradition by the officers and NCOs. After the meal, Mervyn and Frank Richardson decided to go into Calgary town for a look around. When they were at the camp gates, the sentry told them that a Mr Buckmaster and family were expecting two Aussies for Christmas dinner, so the two young fellows accepted this kind offer from a hospitable Canadian family, in anticipation of another good meal in the evening.

But the Buckmasters promptly sat down to another midday meal of roast turkey and all the trimmings. The young Australians heroically ate their share, then went with their hosts on a tour of the neighbourhood homes. In many of these, nobody was home although drinks were set out for callers, and eventually everyone congregated at one of the homes, where the celebrations continued. 'It was a memorable Christmas for me, 1940,' said Shipard.

Training at Calgary was on Avro Ansons, which Mervyn and Frank called 'Angel Makers', taking them as close to Heaven as they would ever go. It was mid-winter, but they survived the climate and enjoyed the stay. The friendship and hospitality of the Canadian people made indelible memories for the transient Australians. Shipard was interviewed for a commission, and confronted the head of the panel, nicknamed 'Smiler' because he never did so. During the interview,

'Smiler' asked what the young Shipard had done before joining the RAAF. 'I was an auctioneer,' he said, to which the question was, 'And what did you sell?' Young Shipard replied, 'Pigs', and watched 'Smiler' throw back his head and laugh.

In the mysterious ways of the Service, Shipard became a Pilot Officer and Richardson a Sergeant when they graduated as pilots. Richardson was recommended for fighters and Shipard for bombers when the course departed on 8 April 1941 for Montreal. The two detoured to visit Niagara, rejoined the others and departed from Halifax for the UK in a convoy on 25 April. The captain of the ship had only two hours' notice that another 100 passengers would be aboard for the voyage. Accommodation was limited, and Ship was allocated a bed in the sick-bay, while Richardson's was a bath.

The ships plodded along at eight knots, and there was little to interest the young aircrews until about the fifth day, when suddenly a cluster of British destroyers raced past from the rear and began depth-charging an area ahead of the convoy. Their ship swung away and, with a destroyer as escort, went to full speed and headed north. Next morning, they were in Rekyjavik. After five days at an RAF transit camp nearby, the remainder of the Atlantic crossing was made on a fast ferry which delivered them to Glasgow.

At the railway station, Ship walked up to look at the puffing engine, and remarked to another of the Aussies that they seemed to have small engines in England. A Scots engineer nearby smartly informed him that, 'Ye're no in England now, laddie! The engines are wee — and powerrrful!'

In due course, the draft arrived at Uxbridge. While there, Pilot Officer Shipard was measured for and received his hand-made tailored officer's uniform from the firm of Jeeves of London. On another occasion, the Australians were invited for tea at Windsor Castle, where the Queen chatted with them in the Crimson Drawing Room, and the young Princesses handed around plates of sandwiches. It was a very different world from that around the country town of Henty.

The draft was informed that they were to undergo a night-vision test. Frank Richardson informed Ship that he intended to fail the test, as he did not want to go onto nightfighters; Ship decided to do his best. On such things are fates decided. Richardson duly failed the test and ended up on bombers, being lost on a raid to Berlin when his aircraft took a flak hit in the open bombbays and exploded. No trace of aircraft or crew was ever found.

Shipard was posted to 54 Operational Training Unit (OTU) at Church Fenton, flying Bristol Blenheims fitted with the new-fangled Airborne Interception (AI) equipment, later and more commonly called radar.

The war came very close to Mervyn Shipard at 54 OTU. On one night flying detail, for which an Airspeed Oxford was to be used, he started the engines and taxied to the holding point, then waited as the engines were too cold for take-off. Another aircraft landed, taxied round and lined up for take-off. Meanwhile, Ship's engines had warmed up sufficiently and as the other aircraft began its take-off roll, Shipard moved out to commence his own. His Oxford had just lifted off into the night sky when two vivid streams of tracer flashed overhead from behind, hit the Oxford in front and blew it to pieces. Ship never saw the German intruder.

When the pilots arrived at Church Fenton, they were teamed ('crewed') with a radio-observer—except Mervyn Shipard. Allocation was made alphabetically, and as he was last on the list, and there were not enough observers, he began training without a permanent radar operator. After two months, and with the course only a short way from completion, he began to fly with the man who was to be with him on all his night combats.

Douggie Oxby was born on 10 June 1920 in Cardiff, Wales. He matriculated but was unable to enter university due to the stringent economic circumstances of the the Great Depression. From 1936 to mid-1940 he was trained and employed as a barrister's clerk, with most of his pay going to support the family. He volunteered as a clerk in the Royal Air Force Volunteer Reserve (RAFVR) to avoid conscription at age 20 into the Army, as Britain was then confronting Germany in Europe. Oxby's bed-ridden mother had insisted that he remain in groundcrew, not aircrew. However, he soon found himself at Yatesbury, Wiltshire, being introduced to the secret equipment later known as radar. He was posted to Wales and the early radar was used to track Luftwaffe aircraft and the occasional surfaced U-boat, but it was a humdrum existence. In 1941, he volunteered for aircrew and arrived at Prestwick, Scotland, for training in the first

Douggie Oxby (left) and Mervyn Shipard with 'Slippery Ship II', 13 victories recorded on the nose, 1943. (D. Oxby)

versions of AI equipment. The course lasted for six weeks and gained him 18 hours flying time — only six actually on the radars.

The young Douggie Oxby, still 'the lowest form of animal life' in the RAF, an Aircraftman 2nd Class (AC2), was then posted to 54 OTU. There, he was 'crewed' with Pilot Officer McMurtrie RNZAF, but before any flying could be done, Oxby went to hospital with mumps, which he believed he contracted from a young Scots lass with whom he had been dallying. At the end of July 1941, Oxby came out of hospital to find McMurtrie had been killed — following ground control, he had flown into a hill.

So Douggie Oxby met Mervyn Shipard and began one of the successful partnerships in the Allied nightfighter force. Their first flight together was on 2 August 1941, in Blenheim IV Z6090. They flew and trained for the next six days and on 8 August, Oxby was qualified as an Observer Radio, and so was deemed suitable for squadron employment. His total flying time was 35 hours 10 minutes, and he was still an AC2.

On 20 August, the newly-qualified crew was posted to an experimental nightfighter unit, 1451 Flight, at Hunston, Hertfordshire, which was using US Douglas Havoc A20 aircraft fitted with a searchlight in the nose, the 'Turbinlite', one of the less successful means of coping with the German night bombers. The idea was that the Havoc would take-off with a Hurricane flying on each side, and when the German bomber was located, the Havoc would illuminate it with the searchlight so that the Hurricanes could shoot it down. 'I could see that the Havoc would receive the full blast of enemy guns,' said Ship. 'I informed the CO I was not particularly happy and would only use the light when coming in to land.'

The same day their postings were changed, to 68 Squadron, commanded by the great Max Aitken. Operations in the area were very quiet, and the new crew converted to the Bristol Beaufighter, as well as flying in co-operation exercises with the ground searchlights and radars. The Shipard/Oxby crew flew and trained intensively in Blenheims and Beaufighters using Airborne Interception (radar) AI Mark IV, and made their first operational sortie on 30 September. It was uneventful, and brought Douggie Oxby's flying hours to 78, with 15 at night.

The early period of nightfighter development has been well described in several excellent books, and little detail of the frustrations experienced by all crews will be given here. However, as with other types of flying, more friendly losses were incurred by bad weather, pilot inexperience or error and equipment failure than by enemy action.

As an AC2, Douggie Oxby had to resort to chits signed by an officer (Shipard) to excuse him from all the menial duties which befell the lowest of the low in the rank structure. The mere fact of being required to fly as an Observer Radio had little standing in the eyes of some of the older peacetime RAF members. ACs were rostered for guard and other duties, and AC2 Oxby was included. Max Aitken had started the administration procedures rolling to have Oxby promoted Sergeant, and in the meantime made up the difference in pay out of his own pocket.

Being the junior crew had its disadvantages, as was soon discovered. 456 Squadron RAAF was in process of converting to Beaufighters, and was based on the Isle of Anglesey, at Valley airfield. As 456 was non-operational at the time, 68 Squadron was tasked to provide one aircraft each night, on detachment to Valley, for the defence of Liverpool. There were few Germans, so the detachment was used as a means of placing junior crews on actual operational patrols, but with the senior crews able to remain where the prospect of action was higher. Pilot Officer Shipard and Sergeant Oxby flew several uneventful patrols from Valley. 'Valley was a very desolate, windswept, sand-blown airfield, with cold winds off the North Sea,' recalled Ship. 'Accommodation was a small tent off the end of the runway, almost under the wing of the Beaufighter. There were two camp stretchers, a kerosene lantern and a telephone. The evening meal was a bowl of hot soup. Douggie and I spent more than our share of time there.'

On 1 November, the Luftwaffe bombed Liverpool, and the Shipard/Oxby crew had their first taste of combat in Beaufighter Mk1 No.7540, armed with four 20mm cannon. After take-off at 20.24, they were patrolling Bardsey Island at 10,000 feet, when ground control gave them a course to steer (vector) of 270 degrees, to an unidentified aircraft — a 'bogey' — at 5,000 feet. They turned after the aircraft, and Oxby soon had a return on his radar: a contact at 7,000 feet, maximum range, over the sea west of Bardsey. Then it was lost, but after more vectors another contact was made, at 11,000 feet and maximum range. This time the aircraft was 500 feet above and 30 degrees to port.

Now Douggie Oxby gave Shipard the course corrections to bring him into position on the enemy. The enemy was climbing, and after throttling back to 110 knots, and lowering 15 degrees flap, Ship saw the silhouette of the German at 1,000 feet range (330m). Cloud was at 5–7,000 feet, but both aircraft were in bright moonlight at 12,000. Shipard closed in from behind and flew directly underneath. He and Oxby looked up at the unique elliptical wing plan-form of the Heinkel He111, but could not see any exhaust flame. Ship closed to 400 feet (130m) and only 10–20 feet (6) below, dead astern, and fired a two-second burst from the four 20mm cannon. The 60 shells struck home and at once the starboard engine caught fire, then the fuselage interior seemed to be ablaze, the Heinkel fell steeply to port and spiralled down into the clouds. The glow could be seen through the clouds, then there was an explosion and fires on the ground as the Heinkel crashed.

The Heinkel had been a H-6 variant, number 430, F8+KR of 7/KG40, flown by Leutnant G. Leins. Examination of the wreckage showed the starboard engine had caught fire, and then the Heinkel began to break up in the air. The port engine was found 800 yards (metres) from the main wreckage, which hit near Gwalchmai, Wales. The explosion reported by Ship was the detonation of a 250kg bomb. None of the crew survived the action, and Leins' body was never found, nor identified in the wreckage, but the remains of the other three—Unteroffizier Tepe, Unteroffizier Fischer and Gefreiter Terstegen—were buried in the German Military Cemetery at Cannock Chase.

'Quite a memorable combat,' recalled Shipard. It had been the first of Douggie Oxby's 36 visual sightings of enemy at night, and the first of 13 aerial victories he would share with Mervyn Shipard. For Oxby, this victory was 'the outstanding thrill of my operational career.'

This victory by a new crew led to an upsurge of interest in the Valley detachment by the older hands, and Ship was not sent back there for two weeks. More uneventful patrols followed and on 27 December Shipard was told he was tour-expired and to go for a rest as an instructor to an OTU. The alternative was to volunteer for overseas service. This they did, with no hesitation, to 89 Squadron, based at Abu Suweir, near Ismailia, Egypt.

On 4 January they flew their last sortie with 68 Squadron, and then reported to No. 1 Ferry Training Unit. They sat through lectures on the ports of call and routes, and received tropical equipment and clothing. On 11 February 1942, with other Beaufighter crews, Shipard and Oxby went to the Bristol factory and took delivery of a brand-new Beaufighter, which had small extra fuel tanks in the engine nacelles. After carrying out some fuel consumption test flights and navigational exercises they began their journey on 17 February 1942 to the Middle East. With a Coastal Command Beaufighter leading, and another nightfighter Beaufighter, they set off for Portreath and next day began the long over-water flight. As Oxby had no navigational training, he was not confident of sucess, but 'Ship was perfectly capable of getting us there, so we landed without incident' at Gibraltar. Ship was not at all worried, as the navigator in the Coastal Beaufighter was doing all the work. Rain delayed them for five days, then they set off across the Mediterranean.

The Beaufighter was totally unarmed, so when they were intercepted near Lampedusa by a gaggle of Italian Fiat CR42 fighters, Douggie was reduced to flashing red Aldis light bursts at the Italians, who did not seem to be worried at all. Ship — who did not actually see the enemy — went down to sea level and out ran the CR42s but, for the first time in a flight with him, Oxby had been frightened.

They were en route to Egypt via Malta, but were to return to the island for a tour of duty.

Before Italy entered the war in June 1940, conventional military and political wisdom had been that Italian air and naval supremacy made it impossible for Britain to hold the island if war with Italy became reality. Consequently, little had been done to provide defences or the supporting system necessary in modern warfare. However, when Winston Churchill replaced the ineffectual Neville Chamberlain as British Prime Minister on 10 May 1940, the concept of Malta's role changed. During the war, and for years after, the dogged defence of Malta, isolated in the middle of the Mediterrranean and only 60 miles (100km) from Italian air bases, was portrayed as an example of British bulldog spirit, courage in the face of adversity, and determination to fight and win. It was crucial to hold Malta, despite all the Italians and Germans could do, because of its importance as a base for gathering signals intelligence from the central and eastern Mediterranean and southern Europe, reaching into southern Russia. This information, part of the 'Ultra' organisation, was worth the hundreds of aircraft, dozens of ships and thousands

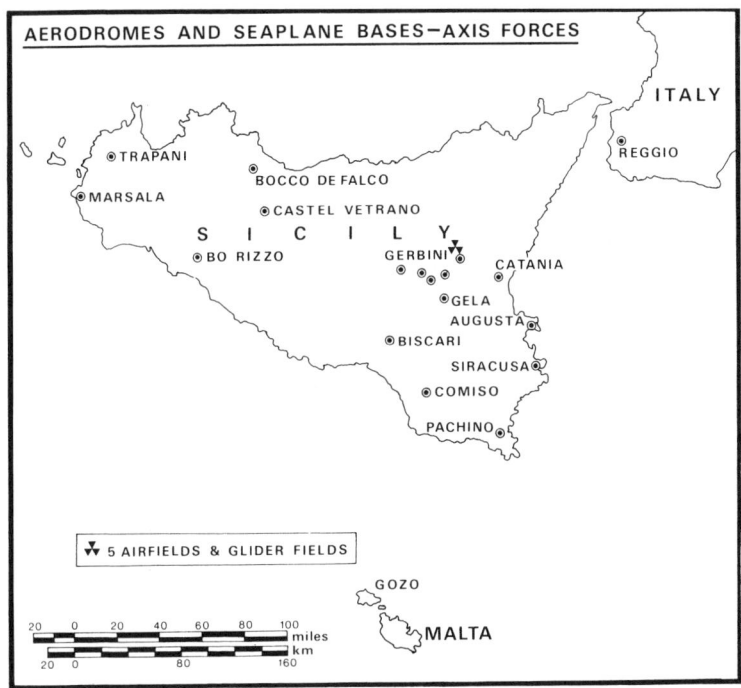

of lives expended to hold the rocky island. The siege and defence of Malta, 1940–42, is one of the epics of warfare, and ranks on the Western side with the determined Soviet defences of Leningrad and Stalingrad. Without the successful defence of Malta, the course of the war in the Mediterranean would have gone against the Allies, and affected the world situation.

At Gibraltar, the crews had been told that enemy fighters circled out of range of the anti-aircraft weapons, waiting to pounce, so it was best to cross the island coast low and fast. 'We certainly crossed low and fast,' said Ship. 'So low and fast we went across the island before we had time to turn. As we got out of the aircraft, the ack-ack opened up and we spent the next 30 minutes in a ditch.'

The Malta authorities had been in the habit of taking for their own use any aircraft which landed in transit and which they decided was in need. The new Beaufighter was attractive to the Malta people, but Douggie pointed out that it had no radar or guns, and was useless to the island. They were allowed to fly on to Africa on 25 February with the other two Beaufighters.

As they neared the coast a swirling dust-storm made landfall difficult. To add to difficulties, the groundcrews at Malta had not filled the newly installed fuel tanks in the nacelles, and the expected 30 minutes additional flying time was found not to be available; they would have to land soon. After following the vaguely seen coastline for some time, a desert airfield was seen below, and Ship led the formation down to land. However, the airfield was only 700 yards (metres) long and he ran off the far end, damaging propellors and landing gear, the Beaufighter ending up on its belly on the sand; the other two got down safely. They found the field was a Fleet Air Arm installation and the RAF visitors were looked after as well as possible. Next morning, Ship was to go on as a passenger in the Coastal Command Beaufighter which was the formation leader. He suggested the pilot hold it with brakes while running up to full power, then release the brakes for the shortest possible take-off. The huge Bristol Hercules roared, brakes on, the pilot released them, the plane surged forward and then in only 30 yards (metres) the powerful torque effect of both propellors swung the Beaufighter to the right, off the strip and into an aircraft pen. The Beau was wrecked. The stone pen was being demolished by Egyptian workers and, tragically, two were killed. The Navy had seen one unsuccessful landing, one unsuccessful take-off and two wrecked planes since the RAF arrived, so sent the transients the rest of the way by train.

'The first few days in the desert were very demoralising. Tents with dirt floors, camp stretchers and kerosene lanterns; Messes were wooden buildings lined with hessian. The redeeming feature was four other RAAF pilots in the squadron: Digger Ross, Red Gray, Chicka Crombie and Spurg Spurgin.' John Ross described Shipard as, 'in my opinion, the keenest pilot I met during the war.'

The role of 89 Squadron was to protect the Suez Canal. Cairo was an open city, and 'lit up like a Christmas tree at night,' recalled Ship. He and Douggie Oxby flew practice interceptions, co-operation flights with the searchlights and radars and other flights to facilitate testing the radars and anti-aircraft guns. Soon after he arrived at Abu Suweir, Ship saw Frank Richardson's name on a list of those missing on operations.

Anzac Day 1942 was a memorable occasion. Four of the five RAAF pilots — Spurgin neither drank nor smoked — had been drinking the day before with members of an Australian Army unit based near Abu Suweir. The Army intended holding a dawn Anzac Day service, but nothing was arranged for the RAAF men with RAF squadrons. 'Just before dawn on 25 April, three Beaufighters departed Abu Suweir', said Mervyn Shipard. 'The first was flown by Red Gray, the second by Chicka Crombie and the third by John Ross.' Shipard did not consider himself recovered from the drinking of the previous day, and went as a passenger in John Ross' aircraft. 'We timed our arrival all very nicely, just when they were lined up for the Dawn Service. Three Beaus at zero feet could be somewhat disconcerting! Crombie claimed one tent destroyed by his slipstream.' The three aircraft criss-crossed the Army camp area, and from his back seat Ship could see soldiers ducking and scattering every which way. Then they returned to Abu Suweir for three less than perfect landings. Crombie slammed his Beau down so hard it bounced a good 50 feet (15m) and at the top of the bounce it could be seen that the pilot's hatch was open and Crombie had both hands clasped over his head in the boxer's victory salute!

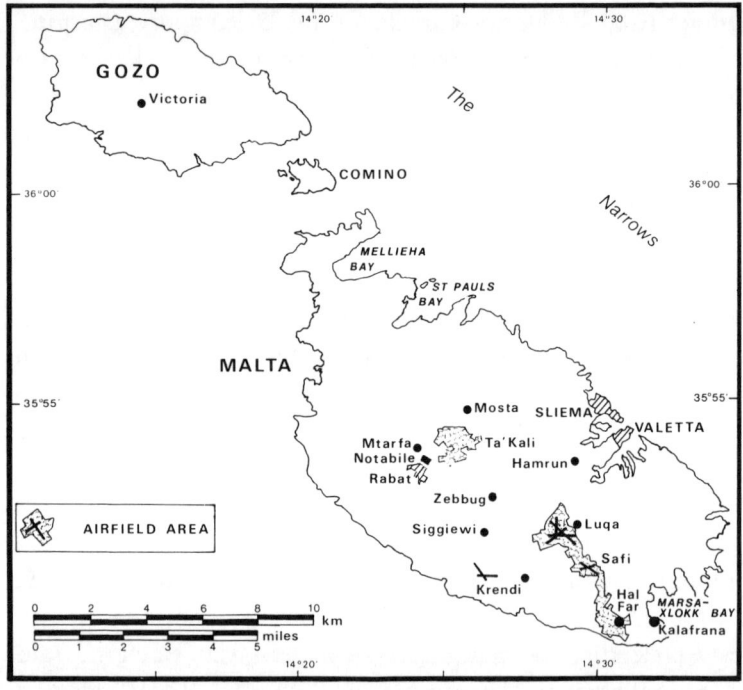

The CO of 89 Squadron, Wing Commander George Stainforth, called for volunteers for a detachment to operate from Malta, which was under heavy attack from the Luftwaffe and Italian Regia Aeronautica. Ship and Oxo were among those who came forward. After four months in Egypt, Ship and Oxby arrived back on Malta on 22 June.

The crews had no illusions about the combat situation, but

their accommodation left something to be desired. It was a bombed block of apartments, with no windows or doors, and the crews slept — or tried to sleep — on camp stretchers, on straw mattresses, on concrete floors. The building looked out on Sliema bay, in peacetime a very desirable location, but as it was under the flight path of attacking bombers, was not now so popular. As the squadron operated at night, and most enemy attacks were in daylight, sleep was disrupted. The crews found the best place to catch a little sleep during the day was on a blanket spread on the rocks beside the bay — here an hour or so could be enjoyed between raids.

Mervyn Shipard adopted his own routine which allowed him to sleep but be ready for take-off in seconds. When on readiness, he would sit in the Beaufighter cockpit, strapped in, with the seat-back lowered to a comfortable position. He would sleep while Douggie Oxby waited in the underground control room, near the telephone. When the order to scramble came, Oxo would run to the aircraft, calling to Ship. With cockpit check already done, eyes instantly adjusted to the darkness so all Ship had to do was raise the seat-back and press the starter buttons for the engines. By the time Oxo was inside and had closed his hatch, 'I would have the engines running and within minutes we'd be airborne. On return, a cup of tea while re-arming and refuelling, then back to the cockpit for shut-eye,' Ship recalled.

On 3 July 1942, despite five contacts, the first Shipard-Oxby patrol from Malta was unsuccessful, as useless radar made it impossible to engage, and two more patrols also brought no result for the same reason. However, on 6 July in aircraft X7642, they were guided close enough for Oxby to take over the contact, and he brought Ship in to engage a Junkers Ju88. Unteroffizier Mellein and crew were lost in M7+MH of KGr806.

Six days later, the harzads of operational flying on Malta

A Junkers Ju88 bomber being shot down. (RAAF official)

provided another experience for the Shipard-Oxby team. X7642 was to be taken on a night-flying test, and Ship did so with an air raid in progress. However, just before lift-off, the port tyre was punctured, probably by shrapnel. Test completed, he touched down, but the heavy plane slewed port and swung around, at speed, tore off the port wheel, the wing ripped away, the fuselage rolled over, the starboard wing tore off, and the fuselage 'rolled about 400 yards (metres) down the runway with Oxo and I inside it.' The only personal damage suffered by either was a broken flying goggle lens on Ship's helmet. They flew again next day, and scrambled on the 14th.

The next successful combat was on 19 July. On this occasion, Ship, in Beaufighter X7702, saw a parachute open as a crew member escaped from the dying Ju88. German records indicate Leutnant Siegfried Sack and crew of B3+PH, I/KG54, were the victims.

Operations of all types were hindered by unserviceable equipment, but the nightfighters, and the vulnerable new airborne radars, were plagued by faults which had a greater effect on their effectiveness than on the dayfighters. On a scramble on 21/22 July, in X7695, their radar failed, so control told them to go to the south of the island and circle until the raid was over. But Douggie Oxby always carried a small torch and a screwdriver with him. He stripped and re-assembled the radar, tested it, and told Ship it was working. Control brought them back into the combat area, they fastened onto a bomber and destroyed it. This time two parachutes were seen, but the loss has not been identified in detail. It may have been Leutnant Leo Skedla and crew in M7+GZ of KG806, who also may have been engaged by 'Moose' Fumerton. Next day, Ship went to the flight commander, related the incident, and nominated Oxo for the Distinguished Flying Medal. Some months later, this DFM was announced.

The nightfighters had to be held on the ground on Malta because of shortage of fuel. The controllers had to wait until the enemy aircraft climbing over Sicily was seen, on radar, to be heading for Malta, and not some other destination — such as Sardinia, Crete or Africa. However, the enemy were at about 21,000 feet and flying at best speed to bomb and get away again. The Beaufighters were at a great disadvantage. Ship wondered how the Bristol Hercules engines withstood the strain. 'We would take-off at full bore, 2900 rpm, at 2000 feet reduce engine revs to 2600rpm, at 8000 feet go to high blower (supercharger), and get to 15 or 16,000 feet, and see the enemy going over our heads 5,000–6,000 feet above, for Malta. No way could you get to them.'

When the enemy had bombed, the pilot would put the nose down and dive back to Sicily. Sometimes Ship could see the exhaust flames, 'four little twinkles. I was about a mile away, and do you think I could catch him? I'd put the Beau down a gentle dive, and when I got beneath him I'd be two or three thousand feet below. I'd ease back (up) and fall behind; this would go on all the way across to Sicily, when I'd have to call off the attack.' The still secret radar was not allowed over the enemy coast. On occasion, when climbing hard to meet the approaching bomber, Ship and Oxo would see, far below on the sea, a quick scattering of twinkling spots, small sparks of light, then the enemy would turn and head back for Sicily. It was assumed that the bomber crew was Italian, and had released their bombs into the sea to avoid hurting relatives on the island. Actually, the crew could have been either Axis nationality, and there are documented cases of Luftwaffe bombers also releasing into the sea well short of Malta. The island had a fearsome reputation among enemy bomber crews.

But sometimes the enemy made a mistake. 'I got one chap because he didn't fly straight. He was weaving from side to side. I could see his exhaust flames. I was just flying straight. You could see his exhausts appear, cross over, then reappear. When he started to come back across, I was flying alongside. If he was flying in a straight line, I couldn't have caught him.'

On one occasion, Ship and Oxo were doing the very basic test allowed by the petrol shortage, and both were dressed only in shorts and desert boots. Ship was warming the engines and Douggie was testing the radar to see if a ground return was received. Suddenly, over the noise of the engines, they heard a stick of bombs exploding. Ship looked out and saw Ju88s attacking, with Spitfires after them. He told Douggie to close his rear hatch, and called Control as he took off. They were sent 'to 20,000 as quick as you can go', and then spent the next hour shivering until told to go home. 'Gosh, it was cold!' recalled Ship.

They had a very close escape during a night raid, when told to scramble. Ship was on the runway, about to push the throttles forward, when Control cancelled the scramble and he was told to return to his pen. The best way to do so was to taxi along the runway to the far end and re-enter the track system which led to the pens. Peering into the blackness, Ship was slowly rolling along the runway when he saw a dark mass ahead. He flashed the landing lights and saw it was the petrol bowser and tractor, abandoned in the middle of the runway. Ship carefully taxied around, went back to the pen, then walked to the bowser. The tractor engine was still warm, and at once he realised where the driver would be—in a nearby big shelter. He went there, and asked for the driver. A Maltese civilian came forward. Ship asked where the bowser was, and received a voluble gesticulatory answer: 'I do not know! I do not know! The bombs were dropping!' Ship was unimpressed, as he would have hit the petrol bowser while travelling at 100 knots.

The Maltese had no love for the Italian and German crews

who attacked them. During one daylight raid, Shipard was at the pen when he saw a German coming down by parachute. He was going to land about 200 yards (metres) away, so Ship began to walk to the spot. But a dozen or so Maltese women appeared from the fields. They were armed with forks, spades and heavy field tools, and began to advance on the German. He pulled a small pistol from his flying boot and pointed it at the menacing group. Ship came up behind him, and asked what he was doing. The German turned, recognised a serviceman and surrendered the pistol. Ship took it, waved it at the women and told them to go away; they did. The relieved German told Ship that he would have been killed.

The besieged island was short of everything. Aviation fuel was in very short supply, as were parts and some types of ammunition. Food also was severely rationed. A diet of bully beef three times a day was all that was possible for long periods of the siege, but Ship found that he could get it down if it was covered in marmalade. Some private enterprise endured, and at a time when eggs in Australia were sixpence (5 cents) a dozen, Ship paid six shillings, twelve times the price, so he could enjoy one a day for breakfast.

On 24 July, Ship and Oxo contacted a Ju87. Following directions from Control, they had descended from clear sky into haze, Douggie called a contact ahead and Ship looked up in time to see a single-engined aircraft turning. He first thought it was a Spitfire, which were sent to patrol lower levels, but it was only 30 yards (metres) away and he recognised the long angular canopy of the Stuka, then saw clearly two people aboard and the other distinctive recognition features of the Junkers. It rolled and dived so steeply that he lost sight of it and the radar could not find it again. It seems this was an Italian Ju87, flown by Tenente Remo Martini, 102 Gruppo Tuffatori, who was hit in the tail. Martini jettisoned his canopy in case he had to bale out, but found the Stuka was controllable and flew back to Gela, Sicily.

On 29 July, Ship and some of 89 Squadron were down at the beach — or the rocky coast which passed for beach — when they saw a bunch of Spitfires wheeling and diving at something on the water. They were surprised to see a Cant Z506B float-plane touch down, a white singlet waving in surrender from the cockpit. A Beaufort crew led by Flight Lieutenant E.T. Strever had been captured by the Italians and were being flown to Italy from Sicily when they overpowered the crew and flew to Malta.

On this day, Malta had food and fuel for only three weeks. The island was under great pressure. To the end of the month, the enemy had flown 3900 sorties over the few targets on Malta, from which the defenders claimed 149-38-140, losing 36 Spitfires. 340 of the total 550 tons dropped had gone down onto the airfields. During the 31 nights there had been 55 alerts, adding to the strain on the population and military defenders.

A great convoy operation was launched and fought its way across the Mediterranean to enable the island to resist longer. This battle, Operation *Pedestal*, was one of the epics of the war, and both sides, British and Italo-German, fought with great tenacity. Thirteen merchantmen and a tanker were assembled for the convoy. The power of land-based aircraft in narrow waters was by this time well and truly acknowledged by the Royal Navy, which provided two battleships, four aircraft carriers, seven cruisers, 24 destroyers, four corvettes, two tow-ships and two tankers. Another aircraft carrier would go out from Gibraltar far enough to launch a batch of Spitfires for Malta, and eight submarines from Malta would take station to intercept the Italian fleet. By the end of the battles, nine merchant ships and four naval ships had been sunk, with four more navy ships damaged. Two Italian submarines were sunk and two cruisers damaged.

The food and ammunition brought in would allow the island to survive that much longer, but the priceless petrol carried in by the badly damaged US tanker *Ohio* really saved Malta. Without the petrol, the aircraft would have been unable to operate, and the island would have been lost. In addition, Axis convoys to Africa would have been able to cross the Mediterranean safely, and Field Marshal Rommel would have received supplies, reinforcements and petrol when he most needed them—at the El Alamein position.

One of the surviving ships was *Brisbane Star*, and Mervyn Shipard was invited aboard by a crew member. There he had the best meal he had eaten for a long time, with fresh bread cooked in the ship's bakery.

After the convoy battle, there was a slackening in some types of enemy activity, and Ship and Oxo had no successful contacts for some weeks. On days off, Ship sometimes went into the Operations Room, standing behind the Controller, watching the activity. Some Controllers seemed a little removed from reality, and Ship overheard one calling a Spitfire pilot some minutes after the pilot radioed he was baling out, to ask if he had baled out yet . . . On another occasion, a Spitfire leader called to say the enemy were not in sight, so the Controller suggested they might be in cloud, at which the Spitfire leader abused him and pointed out there was not a cloud in the sky. Once Ship himself was being shot at by Malta's anti-aircraft defences though he was well above height for such fire. The Controller denied this could be so, and when Ship angrily told him to come on up and look at it, the man replied, 'Don't worry, it's friendly!'

During September, Ship and Oxby were sent after two enemy aircraft, but no combats resulted, and they flew intruder patrols over enemy bases on Sicily. Shipard wondered how much these operations contributed to the war, and presumed they were flown to annoy the enemy.

On 29 September, he decided to attack the enemy motor-torpedo boats at the Porto Empedocle base with the two 250-pound (125 kg) bombs each Beaufighter carried under the wings. It was a brilliant moonlight night, and he went at the boats very fast, dropped his bombs, turned starboard and opened fire with cannon and machineguns. At that instant the enemy anti-aircraft guns fired.

Douggie Oxby recalled it as the most frightening flight of his operational career. 'We were coming from seaward to attack the E-boats, there were cliffs in front of us and Ship had to do a climbing turn away. As we were approaching the wharf along which the boats were moored, we were being fired at from the cliffs, downwards, and from the wharf, upwards. I thought we were bound to be hit, but, I'm happy to say, not a thing.'

'As I roared out to sea at nought feet,' recalled Ship, 'the Bofors shells were coming past on both sides, hitting the sea and ricochetting off into the darkness. I believe Douggie was fairly frightened. He was probably looking backwards, watching the guns fire at us. However, no harm was done (to the Beaufighter).' Ship held the Beau at 100 feet (30m) on the way back to Malta. Radar control called them, and asked if they were returning. Ship decided to see how good the radar was, so did not reply. A few minutes later, the controller said that if he did not reply, another nightfighter would be scrambled to intercept him. Ship duly answered, but believed that he had been on radar, and that no aircraft or ship could approach Malta undetected, even at very low level. They scrambled again on 2 October, but the high-flying raider was too fast and they could not catch him. On 5 October, they bombed the airfield at Castelvetrano, with unknown results.

The constant shortage of all types of goods had led to an

active blackmarket, but the amount of desired items coming onto the island was quite small. The contents of the cargo ships were carefully rationed, and the food for the ships' crews did not go very far. Aircraft on transit to Egypt or Gibraltar and those arriving for duty on Malta were searched and 'goodies' confiscated. One method of smuggling was not detected: 89 Squadron Beaufighters carried cigarettes, whiskey and such in the wing compartments intended for the life-rafts. None of the searchers thought crews flying over the sea would be silly enough to ignore life-saving equipment.

A particularly vicious and uncaring aspect of the blackmarket was stealing parachute silk. The contents of the parachute pack were removed and replaced with newspaper or blankets, and the pack replaced in the aircraft. With so much combat taking place over the sea, the unfortunate men forced to use the parachute simply disappeared with the evidence. It was not until a pilot fell on Malta that the criminal activity was detected, and from then on aircrews had to carry their parachutes instead of leaving them in the aircraft.

As the silk pyjamas, shirts, underwear, etc. were purchased from local shops, it was assumed by many RAF personnel that Maltese were responsible. But the Maltese held the aircrews, particularly fighter pilots, in high regard, and almost every account of the siege includes examples of this feeling. To steal the silk, one would need access to an aircraft, presumably with a vehicle, and not seem out of place around the machine. Whoever it was knew their thieving resulted in deaths of men defending the island. Those responsible were not known to have been caught, but if they had been, given the spirit of the place and times, they would have been lucky to have survived to face British justice in a court.

Douggie Oxby had a brush with the parachute thieves. When in Valletta, he bought a pair of silk pyjamas but on returning to his quarters and examining them more closely, found they had been made from parachute silk. On the very next scramble, as the Beaufighter was climbing for height after take-off and he had tested the radar, he felt around under his seat for the parachute pack. It was missing.

In mid-October, the Luftwaffe began a week-long period of attacks, the last period of intense air battles over Malta in 1942. In May, the defenders barely coped with overwhelming enemy attacks, but after fighter reinforcements during that month, and later, the defences were far more formidable, and the Germans lost aircraft and crews for little result. Ship and Oxo scrambled on 11 October and claimed a Heinkel He111 as probably destroyed. They scrambled again, in V8219, and destroyed a He111, probably 6N+HH of 7/KG100, which failed to return at this time. On 12 October, in Beaufighter X7840, they destroyed another, probably 6N+AH of 5/KG100.

The Luftwaffe returned again on 13 October, and on this occasion happened to pass over the 89 Squadron detachment quarters, an apartment block at Sliema, on the northern side of Valletta Harbour. Crews on night duty were supposed to sleep during the day, but the din of the air attacks, exploding bombs, and particularly the anti-aircraft battery firing from nearby Spinola Point, made this impossible. Off-duty crews therefore went out onto the road or balconies to watch and cheer the Spitfires. But the Ju88s began to release their bombs early and it became obvious that the missiles would arrive close to Sliema, so the spectators stampeded for cover. Ken 'Red' Gray, another RAAF member of the squadron, dived under his bed as the explosions began, and when all was quiet found he had been beaten there by Mervyn Shipard, who from that time was known as 'Slippery Ship'.

The spirit of the times and the men was also demonstrated by Chicka Crombie, who destroyed seven enemy aircraft during his time on Malta. Crombie was returning to the building after breakfast when the raid began, and took cover under the staircase. As the bombs exploded, Crombie suddenly stood, shook his fist at the Ju88s, shouted, 'You'll never get me on my guts again!' and proceeded to walk upstairs.

Ship and Oxo were on duty on 14 October, and scrambled after a fast raider, who beat them back to Sicily. Because the radar aboard was still secret, Ship gave up the chase, and noticed that there was a problem with the starboard engine: the instruments showed no oil pressure and the temperature was 'off the clock'. He turned back for Malta, and when Control called to report another bandit, Ship replied he was on one engine. They were at 15,000 feet. Suddenly, Oxo called that he had a contact, 'coming straight for you, about 4,000 feet below!'

Ship waited for Oxby to call the moment to turn after the enemy, then rolled the black Beaufighter onto its back, both engines throttled back, and saw the German below as they swept down onto its tail. The big fighter accelerated quickly and Oxo called that they were closing too fast, so Ship dropped wheels and flaps and came 'screaming in to this chap.' Oxby, watching from his rear canopy, saw they were only 30 yards (metres) behind the Heinkel. 'When I was almost on him he started to climb,' Shipard continued. 'That's all I wanted; I pulled back on the stick and opened fire. He blew up and we flew through the rubbish; all over the aeroplane. I opened up the other engine, Oxo yelled out that we were on fire, but it was German petrol.'

The Heinkel was probably 6N+EH of 5/KG100. The Luftwaffe fuel quickly burned off, but ignited the oil and grease on the extended Beaufighter undercarriage, but that too soon burned away with no damage. A few panels of fabric burned off the rudder and ailerons of V8219. Ship retracted wheels and flaps and continued to base. For Douggie Oxby, 'that was the most exciting evening of the whole war!'

On 18 October, the pair flew their last interception over Malta. In X7777, they were directed onto a bandit, but it was too fast and fled to Sicily at a true air speed of 480 knots. Ship had made 102 flights from the island, of which 41 were operational. On nine flights unserviceable equipment of one kind or another had removed chances of success. Ship had been happy enough with his detachment to Malta, which he thought 'a good spot. I enjoyed it. The food. I got sick of bully beef three times a day, but apart from that it was a good life. There was always something to do.' Together, Shipard and Oxby had destroyed six German bombers and probably destroyed another.

As for many thousands of other servicemen, their time on Malta left a deep impression. 'Today, if an air raid warning sounds on television, my mind goes back to the island. The dedication and heroism of the people of Malta, military and civilian, will never cease to amaze me. Little wonder the island was awarded the George Cross.'

They flew to Egypt in a DC3, 8½ hours to Cairo. There were 40 people, men, women and children, packed into the fuselage, plus a full load of fuel. Later, Ship often wondered what the gross weight of that aircraft would have been as it began its take-off roll. On arrival, Ship was served 'a breakfast I will never forget. Three helpings of bacon and eggs — four eggs in each helping.'

After the intense activity around Malta, the eastern end of the Mediterranean was a lot quieter. They found themselves on the duty roster from 31 October, and were detached to Palestine for the defence of Haifa and Tel Aviv, but there were no contacts and until the end of November they flew practice sorties. Then on 4 December, the Shipard/Oxby crew was detached to Bu Amud, a temporary airfield established for the defence of Tobruk. General Montgomery's 8th Army was chasing the Italo-German forces west along the

African coast, and Tobruk was again an important port. US-British forces had landed in Morocco and were pushing the Germans there to the east. The Germans continued to fight hard, and Rommel taught the Allied forces in Tunisia some severe lessons.

Douggie Oxby's operational service, particularly the in-flight repair of the radar, resulted in the award of the DFM, announced on 8 December.

On 12 December, they flew their first scramble since leaving Malta. Their aircraft was Beaufighter X8009, neatly

Mervyn Shipard waiting while 'Slippery Ship II' is refuelled in North Africa. (Mervyn Shipard)

adorned on the nose with 'Slippery Ship', painted by the groundcrew in Egypt. The Luftwaffe was attacking Tobruk, and Shipard went to a position towards Crete, freelancing along the probable German approach path. Oxo brought Ship onto two Ju88s, both of which were destroyed. There might have been a third victory, first detected behind them, but the radar contact was broken as they whipped around to get onto his tail, never to be regained. The rest of December 1942 was devoted to uneventful convoy patrols. 'Christmas Eve was quite a memorable occasion,' recalled Ship. 'Around 10 pm it was decided the enemy also were celebrating Christmas, so we all adjourned to the Mess tent.' Some beer had been acquired and stored for the event. At midnight, Douggie Oxby became so high on rum that he climbed the tent pole — but could not get down. The problem was solved by chopping down the pole. The tent gradually subsided and Oxo was gently brought back to earth. Douggie thought his pilot was responsible, but Ship firmly maintains the axe-man was an Aussie named 'Rus' from the Hurricane squadron which protected the port by day.

On 7 January, 'Slippery Ship' was due for a 30-hour inspection, which was almost completed when a sandstorm blew up. The 20mm cannon had been removed, stripped, cleaned and replaced but not harmonised with the sights — that is, it was not known if they were accurately aligned. Because of the flying sand, it was necessary to leave the covers on the guns. The Luftwaffe attacked and the Shipard/Oxby team scrambled.

Oxby brought Ship in for a visual on a He111, but after the cannon fired only a few strikes were seen on the starboard wingtip and the German escaped. Then another He111 was found, they came in to firing range and, well aware of the off-set guns, Shipard decided to aim at the port wingtip in order to hit the fuselage. The rounds hit the port wingtip. The German escaped.

Bad things are said to come in threes. When landing, X8009 went into an uncontrollable swing to starboard, the wheel came off and the faithful aircraft was wrecked. Shipard and Oxby were unscathed. 'That was the end of "Slippery Ship". A disastrous night,' said Shipard. The sting in the tail was that the accident was written off as due to 'Inexperience on type'. Ship had been flying Beaus since mid-1941.

The Germans had been sending over a Ju88 reconnaissance aircraft daily, but at 28,000 feet it was above the operational ceiling of the Beaufighters and also above the anti-aircraft fire from Tobruk. In an effort to catch this German, Ship's Beaufighter was stripped of the drag-inducing camouflage paint, three of the four cannon were removed, and only a reduced load of fuel was in the tanks. 'Slippery Ship II' X8447, squadron letter 'S', was thus a bare metal lightened model. It was decided to retain the radar in case of bad weather, when it could be used to assist in navigating back to base. The aircraft was a new model, with the dihedral tailplane.

On 14 January, when on a night flying test, Shipard and Oxby were warned that the Ju88 was in the area. They went after it but the German was already on the way home. However, at 20,000 feet, 100 miles (160km) off the coast, they could see him some 2,000 feet overhead and only 2 miles (3.5km) ahead. The bright shiny Beaufighter was seen by the alert Luftwaffe crew, and at once a black trail from the Jumo 211 engines showed the pilot had gone to maximum boost on the way to Crete. He also put the nose down slightly and was 'going like a bat out of hell; so were we', recalled Oxby.

Gradually the Beaufighter came closer, and at 21,000 feet, 1,000 feet (300m) range, Shipard tried a long burst with the single 20mm cannon. There were no results, and the chase went on for 60 miles (100km). Ship then realised that ahead was a cumulo-nimbus, the only one in the area, and the

German was diving for the top of the cloud. Oxby switched on the radar. The German disappeared into the cloud and Oxby called for Ship to turn hard starboard and dive; he was sure the German had done so at once inside the cloud, turning onto a reciprocal heading to dodge the Beaufighter.

Ship noticed the air speed indicator (ASI) showing a large loss of speed, so pushed the nose down so violently that 'dirt off the floor came up past me. We were diving in a very high speed spiral dive. I then woke up to the fact that the ASI was inoperable. The altimeter was winding down at a great rate.' For Oxby, in the back, 'suddenly all hell was let loose. We were on our back and in a spiral dive, which Ship corrected at 11,000 feet, after losing 11,000 feet,' recalled Oxby. Meanwhile, he had been trying to open one of the hatches in case he had to bale out, but the high G-forces in the spiralling Beaufighter prevented him reaching either top or bottom hatch, and he accidentally opened his parachute. So, 'surrounded by silk and somewhat shaken, we limped home, feeling the fates had been playing games with us,' he said. After landing without any further sighting of the Ju88, it was found that the pitot tube must have been blocked by ice as the heater was ineffective at that height, though it warmed up on the ground.

On 16 January, the Shipard/Oxby team scored again in X8447, 'Slippery Ship II'. Flight Sergeant R.G. Peters, with Flight Sergeant F. Halliday as radar operator, was first off, and as Ship was climbing after take-off, there was a large explosion to starboard. 'Peters has got rid of that fellow,' Ship called to Oxo. A contact was gained on radar, but Oxo yelled that they were too high and too fast. Ship looked out and saw the Ju88 below. He dropped wheels and flaps, dived behind the slow-flying Junkers and weaved as he increased speed, until he was behind and slightly below, closed the range, eased his nose up, and opened fire on the bomber. The 88 blew up at once, but the recoil slowed the Beau to stalling speed, it rolled onto its back and spun. 'A spin at night, with wheels and flaps down, was a little disconcerting,' was Ship's understatement.

He recovered and resumed patrolling. 15 minutes later, another Ju88 was picked up on radar, but this time Ship was aware of what they were attempting. Instead of flying at 150 knots, they were plodding along at 110, which was quite slow for the nightfighter and made it difficult for the fighter pilot. Ship slowed down, came up behind the 88 and shot it down. It fell away into the night, and Ship returned to Bu Amud. German records indicate two Ju88s were lost from Lehrgeschwader 1 in the Tobruk area, and 6/KG100 lost 6N+DR, Werk No.7949, flown by Leutnant G. Barett.

After landing, Ship waited for Peters to return, and stood watching the night sky, hoping to hear the sound of the Beaufighter. Later it was found that Peters, when attacking his Ju88, had closed too quickly and flown into the tail of the Junkers. However, as Shipard waited, an aircraft did enter the airfield area, with navigation lights on. Ship could tell by the engine noise that it was not enemy, but also was not a Beaufighter. Quickly he ran out and lit the rudimentary flarepath — a row of ten tins half-filled with sand, into which a little petrol was poured and lit. He stood at the end of the runway and watched the aircraft turn in to land.

It switched on its landing lights, and Ship realised that it was a four-engined bomber. A huge B24 Liberator, with its 110 foot (33.5m) wingspan, roared down the strip, the blast from its outboard engine blowing out the flarepath. Then another, and another arrived, and soon eight B24s were around the small field. The first had run off the runway into a patch of camel thorn, and the Bu Amud fire-truck rushed out to the nose of the bomber in case of need. But the big B24 opened up its four Pratt & Whitney 14-cylinder radial engines and advanced on the truck, which quickly moved away, followed by the B24, and no matter where the fire-truck went, the bomber pursued it. After about two complete circuits of the airfield, the truck escaped. Later it was found that the standard procedure at the B24 base was for a truck to go to the nose of each bomber, then guide it to the correct parking bay . . . The Americans came from a nearby airfield, but decided to stay the night, slept in and around their aircraft, and took off for their own base next morning.

One B24 turned back, landed, and the pilot told Ship that one of the Ju88s he had shot down the night before was only a few miles away along the road to the west. Ship politely thanked the American, but thought he was referring to an old wreck which had been out there for weeks or months. Then 'Rus', the Hurricane pilot, landed and described a new Ju88 wreck along the road. Ship went for a look and, sure enough, it was his victory of the night before, landed wheels up, remarkably intact.

This Junkers Ju88 was L1+RH of 1 Staffel Lehrgeschwader 1, an elite unit, and was the Ju88A-14 model, fitted with a 20mm MG FF cannon in the nose for anti-shipping strikes, plus Kuto-Nase balloon-cable cutters in the wings. It was flown by Oberfeldwebel Herbert Isachsen, on his 328th operational sortie. The Germans had been taken prisoner, and the victorious crew acquired a pistol each from the booty. Isachsen had been on operations since the beginning of the war, over Poland, Norway, France, Britain and the Mediterranean. The award of the Knight's Cross to Isachsen was announced on 3 September 1943.

However, this was the last combat for many weeks. A senior officer from Cairo came to visit, and for the necessary parade, all were to wear more formal uniform — no shorts were allowed on parade. The officer told Flying Officer Shipard that his pilot's wings were sewn onto his tunic higher

Ju88 1L+RH of Lehrgeschwader 1, shot down by Mervyn Shipard, in the desert near Tobruk, 17 January 1943. Oberfeldwebel Isachsen, Knight's Cross, successfully landed the Ju88 and the crew became prisoners. (Mervyn Shipard)

than prescribed in regulations. Ship replied, 'Sir, one day I hope to receive some ribbons and I won't have to have the wings repositioned.' The visitor said, in turn, 'I'll have to see what I can do for you.' Soon after, and it surely must have been coincidence, Douggie Oxby was informed that he had been awarded a bar to his DFM, and also had been commissioned with effect from the date of his interview in Cairo on 31 October 1942. Mervyn Shipard was awarded the DFC, and a week later the bar to it was announced. So far, the combined talents of each man had resulted in 13 enemy destroyed, one probably destroyed and two damaged.

On 22 February, the squadron moved to Castel Benito, and more patrols followed, but only one combat resulted. On 2 March 1943, at dusk, two Ju88s were seen going in to bomb Tripoli. Shipard went in and fired two bursts at the closer Junkers, and both jettisoned their bombs, then replied with machinegun fire. They split up, dived away to left and right and Ship lost sight of them against the darkening sea. He went down to low-level, but no further contact was made. The Junkers was claimed as a probable, but one small calibre hit was scored by the German gunners, right through a lower cylinder of the starboard engine. It began to vibrate and smoke, so Ship shut it down. There seemed to be no effect on handling and Beaufighter X8447 landed safely on one engine. This was the only hit made by the enemy on any of the aircraft flown by Ship during his operational career.

On 18 June, after patrols and convoy escorts with no further action, both men were tour-expired. Shipard did not realise that it would be the last operational flight they would make together. Ship had actively sought as much flying as possible in the past three months, as he said, 'anytime, anywhere'. The faithful Oxby went along on all the flights. They flew 55 hours in March, 68 in April and 36 in May, as they flew in any type of aircraft available to them. The

African campaign had concluded on 8 May, with the total surrender of Axis forces in Africa. The Allies were planning the invasion of Sicily, and the further move to Italy. During their time together, Mervyn Shipard and Douggie Oxby had destroyed eight Ju88s, five He111s, probably destroyed a further one of each type, and damaged two Heinkels. They had been on operations for almost two years continuously.

The RAF was exceedingly careful in its definitions of individual aircrew quality, reflected in the assessments noted in personal flying log books. That this was no ordinary team is demonstrated by those log book annotations. Mervyn Shipard, at 5FTS Calgary, was 'Above Average' in all categories; at 54 OTU, 'Above Average as a Nightfighter Pilot'; on 68 Squadron, 'Above Average as a Nightfighter Pilot', 'an exceptionally keen pilot with 1 He111 so far to his credit'; on 89 Squadron, 'Exceptional as a Nightfighter Pilot'; 54 OTU, 'Exceptional as a Nightfighter Pilot Instructor', 'Exceptional as an Air Gunnery Instructor'. His commander's assessment of Douggie Oxby's abilities as a radar operator also was 'Exceptional'. Mervyn Shipard's comment is, 'Douggie Oxby was a very keen and efficient night fighter radar operator. During our two years together we developed a very strong relationship which still exists to this day.'

'It all came to an end in the middle of June 1943,' recalled Ship. He was told to report to his CO, who informed him that his time with 89 Squadron was over. He was offered a choice: a desk job in Cairo headquarters, a flying job in Ferry Command, or return to the UK. Both men decided to return to the UK. They returned together to the UK by way of West Africa, Lagos, Lisbon and Southern Ireland. Both were in identical civilian clothes, with passports giving their occupation as 'government officials'. Despite questioning by officers of Irish Intelligence, they were allowed to take the BOAC (today's British Airways) flight to Poole, Dorset on 27 July.

Ship went to RAAF headquarters in London, for an interview with Air Vice Marshal Wrigley, RAAF commander in the UK. Wrigley told Ship that he was to go on six months' rest to an OTU, then take command of 456 Squadron RAAF, a nightfighter unit. Ship asked if the six months could be spent on Spitfires and dayfighting instead of at an OTU, but this request was denied, and on 1 August 1943, he returned to 54 OTU, with which he had trained two years before.

54 OTU was now located at Charter Hall, Berwickshire, with a Gunnery Flight at the satellite field of Winfield. After a week, Ship was asked if he would take over this gunnery flight, but warned that first he would have to do a conversion to Spitfires and attend a course on Spitfire gunnery. He accepted at once, not realising that it was to mean the end of operational flying. During his four weeks at Sutton Bridge, Ship was visited by a senior RAAF officer from Melbourne. This officer was looking for experienced pilots to establish an RAAF gunnery school in Australia. He asked Ship if he was interested, but the reply was negative. Ship was thinking of the promised command of 456 Squadron RAAF in a few months.

During October, Ship appeared on the routine Monday morning CO's Parade in his drab RAAF summer uniform, the sole person so dressed among the ranks of blue-grey RAF. When queried, Ship explained that as an RAAF officer he was bound by regulation to wear the appropriate uniform for the season—summer in Australia. 'Get to Hell and never appear on my CO's Parades!' was the response. But for Ship, 'that suited me right down to the ground.'

Meanwhile, Douggie Oxby had gone to Ouston in Northumbria, to train others in the art of nightfighting. The last meeting during the war for them was when Shipard was Best Man at Oxby's wedding in Cardiff on 5 November 1943.

Ship was invited to attend afternoon tea at Buckingham

Palace at 3.15 pm on 1 December 1943, but received the invitation only the day before. His CO was understanding and told him to take a Beaufighter, which he flew to Northolt despite a severe attack of 'flu. Ship took a room at the Regent Palace Hotel, 'spruced up' and attended afternoon tea. Next day he reported to RAAF headquarters, to be informed that he was required again at Buckingham Palace on 7 December, to receive his DFC and Bar from King George VI. The ever-patient CO at 54 OTU told Ship to stay in London until after the investiture, then go back to duty. But on 7 December, Ship was informed on arrival at Buckingham Palace that the King was in bed with the 'flu, and the investiture was to be performed by the Duke of Gloucester. Ship returned to Winfield in the afternoon, but ever since has wondered if he was responsible for passing the 'flu to his King.

In January 1944, Ship was informed that RAAF HQ in Melbourne had signalled for his return to Australia. He tried everything he knew to avoid this but to no avail and found himself on a ship out of Liverpool to New York, then a train to San Francisco. As soon as the ship left Liverpool its canteen opened and Ship bought three one-pound (500kg) blocks of chocolate, went to his cabin, ate the lot, and has not cared much for chocolate since. Thirteen RAAF officers were aboard the US Liberty ship from the USA to Australia and they were not happy to find that their accommodation was simply two large crates, each fitted with eight bunks, but no doors or windows, fastened to the aft deck by cables. Protest brought no change. However, the four-week voyage at ten knots was enjoyable. Dress was bathing trunks, there were no parades or duties, life-boat drills or such, and the time was spent in the sun on deck. From San Francisco to Brisbane, they had no sight of land or other human presence on the sea or in the sky.

Ship was given two weeks' leave and went home to Henty, where he became engaged to his sweetheart, Marjorie Wood, of Albury. He then reported to Air Board in Melbourne, where he was interviewed by the officer who had arrived at the Central Gunnery School looking for gunnery instructors. Ship was told he would be going to 2 OTU to convert to Kittyhawks, then to the Gunnery School at Cressy. Ship voiced his disappointment at only two weeks' leave after 3½ years overseas on active service, and was told that 'we have a war on our hands out here!' He later realised that, in his innocence, he had not understood that to some his five months on Malta was an island holiday, and the Pacific crossing on the stern of a cargo vessel was the equivalent of a holiday on a cruise ship.

RAAF mainland bureaucracy ruled, and because Ship had enlisted in the RAAF in Sydney, he then had to go there to be issued with a flying helmet and flying suit. This required an overnight journey, and he then went to Mildura and 2 OTU, where he converted to Kittyhawks in four days. It was off to Cressy and the Gunnery School for a somewhat disgruntled Flight Lieutenant. On his first flight, Ship spent 45 minutes familiarising himself with the area, then gave a 15 minute display of aerobatics over the air base. Later, in the Mess, the CO told him that he would kill himself if he continued to fly like that, and the CO did not want accidents on his RAAF Base. Ship's reply was, 'Sir, it will be my neck, not yours.'

Soon after, two more kindred spirits arrived. Jack Storey and K.A.S. (Kas) Mann were single-engined fighter pilots from the Burma theatre. Storey's career is described elsewhere in this book. Between the three of them, the gunnery course was organised and began instruction to young pilots, few of whom were to be given the chance to engage Japanese aircraft.

On 9 August 1944 Mervyn and Marjorie were married, and enjoyed two weeks' honeymoon. Three days after returning to duty, Ship broke his ankle playing football for an RAAF team, so spent the next six weeks hobbling along on crutches in Albury. At about this time, he was informed that pilots with two tours of operations could apply to resign if they intended joining an airline. Ship could not see much future in the RAAF, so applied to join Australian National Airlines (ANA) and was accepted. However, the RAAF retained him until his ankle healed, and it was not until December that his resignation was approved and took effect. After 4½ years' RAAF service, Mervyn Shipard had flown 1,169 hours, destroyed 13 enemy aircraft, probably destroyed two and damaged two more, all in the air. He was the RAAF's top-scoring nightfighter pilot in aerial victories.

Shipard completed a two-week ground course with ANA, and made his first flight as a co-pilot on a DC2 on 19 December 1944. He had asked to be based in Sydney, and this request was granted. After six weeks as co-pilot, he was rated as a Captain on DC2 aircraft and on 14 February flew Sydney-Brisbane and return. Ship noted that it was easy to tell the difference between the early hand-built DC3s and the wartime mass-produced machines. The early ones were a delight to fly as they had been well-built, carefully assembled and trimmed, but the production line aircraft had not received such care, and this was obvious to the pilots. Ship's ANA career lasted for 12 years and during that time he flew all the Douglas aircraft used by the airline—DC2, -3, -4, -5, -6, -6B—in Australia, across the Pacific to New Zealand, Fiji, Honolulu and Vancouver, as well as on the Sydney-London route.

At the end of 1957, ANA was taken over by Ansett Airlines, and Ship was told that if he wanted promotion in the new airline, he would have to move to Melbourne. For family reasons, this was not acceptable, so he applied to fly for Qantas and in November 1957 Shipard joined that airline with 13,600 hours. For the next 15½ years, Ship flew Super Constellations, Electras and Boeing B707s on all Qantas' international routes except London-New York. Qantas retirement date for pilots was on 30 June after their 55th birthday, so Ship was only three weeks short of his 56th birthday when he completed his service with the airline. His last flight as a crew member was on 27 June 1973, returning from Auckland with Marjorie and son John aboard. The B707 was escorted from the runway to the terminal by three fire engines. Ship had completed 33 years in aviation, with 23,000 hours on 35 different types of aircraft.

Ship's most frightening flying experience came with Qantas, as senior check captain operating out of Tullamarine. The pilot was to be tested in procedures for coping with engine failure on take-off, and as the B707 passed V1 — the speed at which the take-off *must* proceed, and the aircraft come around for a landing if there is an emergency — Ship brought back the throttle on No. 4 engine to idle, the pilot applied left rudder to hold the big jet on course, they lifted off and were at 200 feet when there was a bang from the tail. The Boeing lost all rudder control, began to yaw and roll violently right, was held at 35 degrees right but continued to yaw right, despite all the pilot could do. The B707 was obeying the immutable laws of aerodynamics and Isaac Newton, rolling inexorably onto its back just out of Tullamarine. Ship opened the throttle on No. 4 engine and the added thrust brought up the wing, they completed the circuit and landed. A rudder hinge had fractured at exactly the wrong moment. A worldwide check of Boeing B707s found others with metal fatigue cracks, and the parts were quickly replaced. Another B707 had crashed overseas, but the cause was not identified until this incident at Tullamarine. If the fracture had occurred earlier, the wingtip which dropped would have hit the ground, resulting in destruction of the aircraft and all aboard. Marjorie recalled Ship was still shaken when he arrived home that afternoon.

Ship now lives in Sydney, devoting himself to gardening

and making home-brewed beer. He also owns a yellow Camaro which attracts a certain amount of attention and generates enough sheer power to allow it to draw away effortlessly from younger men in younger cars who like to rev their motors when stopped at traffic lights. When Marjorie is at the wheel, surprised drivers pull alongside the elderly lady and call out, asking if she wants to sell the car.

Douggie Oxby returned to operations in March 1944, to 219 Squadron at Woodvale in Lancashire. 219 was converting to the De Havilland Mosquito, and the operators also had to master the new US radar called SCR720, or AI Mark X in RAF terminology. As the squadron became more proficient, it was moved closer to likely scenes of action and entered the campaign against the V1 bomb, or Doodlebug as it was commonly called, from Bradwell Bay. Douggie Oxby was flying with Squadron Leader Gordon Merrifield, and on 20 June they destroyed two V1s.

On 30 June, 219 Squadron began beach-head patrols over the Normandy battle area, but the only aircraft seen were friendly. In August, Merrifield went on leave, as did Squadron Leader Ellis' navigator, so Ellis and Oxby flew together. On one occasion, they intercepted and identified 26 US B25 Mitchell bombers. Then the 27th contact was identified as a Messerschmitt Bf110, a Luftwaffe nightfighter. Douggie Oxby had long wanted to destroy a nightfighter, and thought that this was to be the first opportunity. But the Mitchell and 110 both were twin-engined, twin-tailed aircraft, and Ellis decided to pull up alongside the target to get a better view of the distinctive long cockpit canopy of the Messerschmitt. As he flew up alongside, the German must have seen the Mosquito and broke away into the darkness—lost.

Oxby then began to fly with Wing Commander Wilfrith 'Peter' Green DFC, the squadron commander, as Merrifield had been posted to India. 219 Squadron now was at Hunsdon, patrolling the Allied battle front in Holland and Belgium. On 23 September, in support of a Lancaster raid, they destroyed a Messerschmitt Bf110 north-east of Cologne; Oxby's ambition had been realised. Still operating from England, the squadron patrolled the battle area. On 2 October, the Green/Oxby team destroyed three Junkers Ju87s in the Nijmegen area. The Germans had formed night battle units employing ground-attack aircraft to harass the Allied positions, and the slow Luftwaffe aircraft were a problem for the sleek fast Mosquito. Peter Green received the DSO and Douggie Oxby the DFC for this sortie. Green's score now was nine, and Oxby had participated in 17 successful combats.

219 then moved to the Continent, based at Amiens-Glisy. On 12 December, Oxby manoeuvred Green into position on a Ju88, Green slid into an ideal firing position, but an electrical fault allowed the unsuspecting German crew to escape — the guns would not fire. 'Boy, were we mad!' said Oxby. The patrols and combats went on: 22 December, one Ju88 destroyed; 26 December, one Ju87 destroyed; 1 February, one Ju88 destroyed; 24 February, one Ju87 destroyed. Green's score now was 14, and Oxby had participated in 22 victories. Douggie Oxby's last flight with Green was on 28 February. Next day, testing a rogue aircraft, Green was attempting a dead-stick, no wheels, no engine landing, which he made successfully but he was killed by the whiplash effect to his neck.

Douggie Oxby was awarded the DSO, and sent on rest to 62 OTU with a rating of 'Exceptional'. He remained in the RAF, being awarded a permanent commission after the war, and completed many specialist and staff courses, eventually becoming the first Navigator to hold the post of OC Flying at a fighter airfield, at the Central Fighter Establishment, in 1957. With Air Commodore Hughie Edwards VC, Oxby visited Australia in 1959 to interest the RAAF in the British TSR2 and P1154 aircraft, and Type 80 radar. He had a reunion with Mervyn Shipard in Sydney.

During a posting to the British High commission in Canada, Oxby decided he liked life there so much he decided to emigrate, so after three years in Europe resigned from the RAF and began working for the Ontario Government Ministry of Health, becoming Chief of Management Development Section and Assistant Director of Personnel Branch. He retired from this second successful career in 1984 and now devotes his time to the garden, spending the winter months away from cold Canada in warmer Florida.

In February-March 1990, Douggie returned to Australia and stayed with Mervyn Shipard. They refought the Second World War, drank gallons of home brew and 'wept over the scores we might have had if everything worked OK.'

Appendix

Victories by Mervyn Shipard and Douggie Oxby, 68 and 89 Squadrons RAF. All were in Bristol Beaufighters.

Date	Aircraft number	Detail
1 November 1941	7540	He111H-6 destroyed F8+KR 7/KG40 Ltn G. Leins
6 July 1942	X7642	Ju88 destroyed M7+MH KGr806 Uffz Mellein
19 July 1942	X7702	Ju88 destroyed B3+PH I/KG54 Ltn. S. Sack

Mervyn Shipard DFC

21 July 1942	X7695	Ju88 destroyed probably M7+GZ KGr806 Ltn L. Skedla
11 October 1942	V8219	He111 probable He111 destroyed probably 6N+HH 7/KG100
12 October 1942	X7840	He111 destroyed probably 6N+AH 5/KG100
14 October 1942	V8219	He111 destroyed probably 6N+EH 5/KG100
12 December 1942	X8009	two Ju88
16 January 1943	X8447	two Ju88 L1+RH LG1 Ofw H. Isachsen RK
2 March 1943	X8447	Ju88 probable

Take that off my tax!

Arafura Beaufighter

Sqn Ldr Reginald Gordon DFC*

SIX ACES

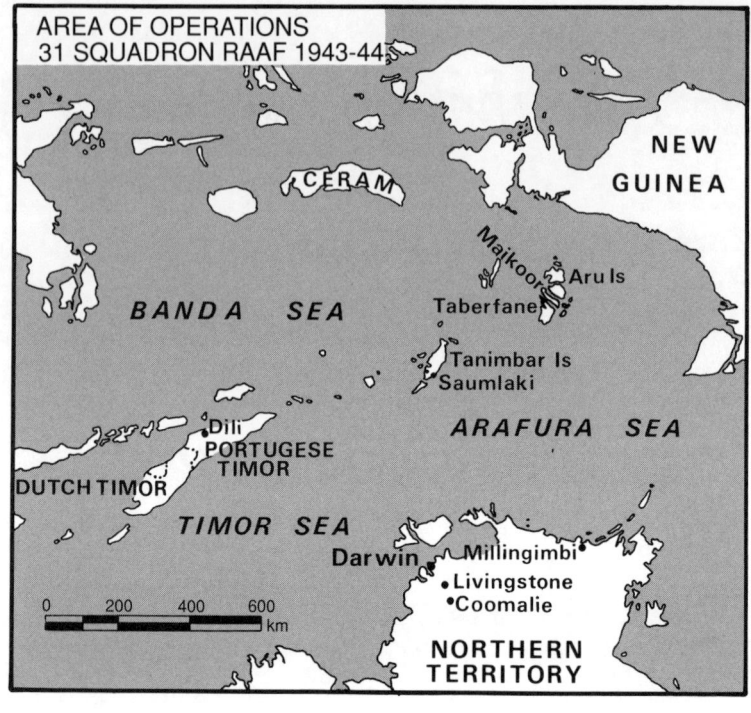

Reginald Gordon DFC

Born in Sydney on 6 July 1917, Reg Gordon joined the RAAF on 16 January 1939, and was allocated service number 485. His civilian occupation was as a clerk — in those days, a position which had status. He successfully graduated from 1 Flying Training School, Point Cook, on 23 October 1939, and was subsequently posted as a Pilot Officer to 22 Squadron which operated the twin-engined Avro Anson at Richmond. Gordon became known as 'Butch'. On 17 June 1940, Gordon and a group of other pilots reported to the Central Flying School, Camden, to undertake a course which would qualify him as an instructor.

On 8 July, Flying Officer F.G. Tampion was demonstrating to Gordon — in Avro Trainer A6-1 — how a slow roll should be executed. During the manoeuvre, Gordon fell out of the rear cockpit, but landed safely by parachute, despite it catching on the fuselage and tearing. None the worse for wear, Gordon successfully completed No.2 Service Flying Instructor's Course on 6 September, and was posted to 1 Service Flying Training School (1SFTS) at Point Cook. He arrived there on 16 September, and began work. On 17 November, in Hawker Demon A1-39, Gordon was involved in an accident. When taxying, he turned to avoid some concrete pipes near the taxy way, but hit some other pipes hidden in long grass on the flying field side of the taxy way, standing the Demon on its nose. Just over two weeks later, in the same aircraft, Gordon had LAC Meecham as pupil. The Demon landed, but at the end of the landing roll the undercarriage collapsed and again the Demon went up on its nose. Investigation showed that the bolt attaching the undercarriage bracing wire to the starboard wing had pulled and sheared.

On 21 December 1940, Gordon was posted as an instructor to 3 Service Flying Training School at Amberley, and there he flew through 1941 and early 1942. When 3SFTS was disbanded Butch Gordon returned to 1 SFTS at Point Cook on 6 April

1942. On 14 May, he arrived at 24 Squadron, Townsville. The squadron was equipped with Wirraway Mk2 aircraft, and if the Battle of the Coral Sea had resulted in a Japanese victory and the invasion of Port Moresby had proceeded, presumably the unfortunate crews would have been used in action to support army units. As it was, the squadron moved south to Bankstown while Gordon was on a course at the School of Army Co-operation and he rejoined 24 Squadron on 28 October 1942. After only a few days, he was with 5 Operational Training Unit (5OTU), flying Beaufighters at Forrest Hill.[1]

During his time at 5OTU, from 2 November 1942 to 19 July 1943, Gordon became known for his daring and flying skill, and on 20 May 1943 he was promoted Squadron Leader. He also gained a reputation as a hard man among the students; indeed, he was not personally liked and when he later went to an operational squadron, some of his former

Butch Gordon, centre front. (Fred Cassidy)

pupils told his navigator, Ron Jordan, that they hoped he'd crash, adding, 'no harm to you, Jordan, but . . .'. However, in the different life of an operational squadron, the dislike faded and Butch Gordon developed a keen and loyal following in 31 Squadron after he began operational flying.

Gordon's long-time navigator and flying companion was Ron Jordan — a combination which became known as 'the old firm of Gordon and Jordan.' Ron had been a butcher, as well as an enthusiastic member of the Nambucca Surf Club, with which he won a bronze medal, and also belonged to the Methodist Young People's organisation. When war was declared, followed by the announcement of the formation of the Empire Air Training Scheme (EATS), Ron volunteered at the age of 17. He needed to improve his educational qualifications, so he studied by correspondence. By the time he was 18, he was accepted by the RAAF, completing courses in

Ron Jordan, observer for Butch Gordon. (Ron Jordan)

radio and air gunnery and then the more difficult navigator-observer's course. Initially, he had been training with a pilot named Peter Gunston, but at 5OTU Jordon was told that Butch Gordon had selected him as his own navigator. Before this, Ron had met Butch only once — during a flight at Wagga — but presumably his prowess had become known. Jordan was the first EATS graduate to go to 31 Squadron, but was soon accepted. They arrived at 31 Squadron on 21 July 1943, to begin operational flying career, and on 1 August Gordon was appointed B Flight commander.

31 Squadron had been raised on 14 August 1942 at Forest Hill, near Wagga Wagga and was moved to the Darwin area on 27 October, before becoming operational at Coomalie on 13 November. Soon after, the Beaufighters began using a strip at Millingimbi, further east along the northern coast of Arnhem Land, which enabled them to reach distant Japanese targets. However, conditions there were even more primitive than at Coomalie. By the time Butch Gordon and Ron Jordan arrived, the squadron had been quite active in attack all across the arc of Japanese-held islands within range, or in defence of RAAF bases in the Northern Territory. Wing Commander Read, nick-named 'Coomalie Charlie', had initially commanded 31 Squadron but on 13 September he handed over to Wing Commander W.F.B. Mann. Read had been in command for a year, having previously flown with 30 Squadron in New Guinea.

In July 1943, 31 Squadron was based at Coomalie, south of Darwin, under primitive living conditions. The buildings had bamboo walls — more or less open — and iron roofs. Trenches were close by, and were used when Japanese bombing and strafing commenced. The squadron's main roles were to cover Allied naval operations, support the Australian commandos operating on Timor, and attack Japanese shipping in the area. Other targets, such as airfields and camps, anti-aircraft positions and supply dumps also were attacked. Japanese air superiority at one time was such that no flights were made to Timor itself, but it was Butch Gordon who recommenced successful sweeps over the area and in so doing raised squadron morale.

Flight Sergeant Tom Warren knew Gordon well, despite the disparity in ranks — 'a thing that didn't seem to worry Butch,' Tom recalled. He added that Gordon 'was a great man, respected and admired by both aircrew and groundstaff for his consideration and great flying and fighting ability. He once told me fear was a feeling he had not experienced and I believe him, judging by the way he would go into the attack irrespective of the odds against him.'[2]

But Gordon was intolerant of fools or incompetents, and those who did not meet his standards received a good tongue-lashing. One of his requirements was for flying at tree-top level, and anyone who persisted in flying a little higher was asked if they were 'up in an aerial circus.' Later, when temporarily commanding 31 Squadron in the absence of the CO, Gordon became unpopular with some of the officers when he nominated NCOs as flight or element commanders on a strike — a decision based on the operational experience and ability of the NCOs, rather than on considerations of rank.

The best operating height for Beaufighters in the areas north of Darwin was ten or twelve feet, or as low as possible. Low flying at speed is dangerous, and inevitably there were losses in training, when aircraft hit trees, or the ground, or when height over the sea was misjudged — particularly if it was calm and glassy, and when the training technique of shadow-shooting (firing at an aircraft's shadow to perfect estimation of deflection) was practised. This low-flying technique was conceived by the successful Australian pilot Clive

Caldwell, when flying over the North African sand dunes, and was later used by many air forces.

Food was not of a high standard at Coomalie, and almost everything came in tins. Techniques of food preservation, dehydration and reconstitution were in their infancy, and when combined with the problems of supply across the great distances from the factories in south-eastern Australia to the far north, some understanding may be gained of the reasons for poor quality and repetitive food. However, such considerations were no comfort to the members of the units stationed in the tropics who had to eat what was prepared by the cooks. Aircrew were given red meat twice a week in the rations, but in 31 Squadron this ration was pooled and shared with the other members of the unit. Sometimes feral pigs were caught and penned, fattened, then slaughtered, to add variety to the food. The beer ration was two bottles per man per month.

'Life at Coomalie Creek was just an existence,' recalled Tom Warren. 'Food was only bully beef, et cetera, dehydrated potatoes, tinned butter — it stank — or smelly margarine. No beer until about July 1943. Nowhere to go except to outdoor movies once a week. Officially we had a stand-down day once a week, but that was a laugh, as we aircrew had to be available as required.'[3]

Aircrew operating over the hostile areas north-west of Australia did so in the knowledge that as well as the implacable enemy and the sea, there was almost no rescue resource provided. Catalinas operating from Darwin had been forced by Japanese raids to move away and were based in Townsville. It took about 24 hours to get a Catalina from Townsville to Darwin, were one to be requested. The only other amphibian or flying boat available was a Walrus, which cruised at 90 knots, and had a range of 400 nautical miles (640km). 31

Squadron operated against Timor and other islands some 520 miles (840km) distant. In early 1943 these were the longest operational over-water flights in any of the theatres of war. Once the aircraft left Australia, they were over water until they reached enemy-held islands. Unlike in Europe, there were no navigational aids for the crews. There was one radio base in Darwin, but this was partly negated by 'dead spots' between there and Taberfane on Aru Island, north-east of Darwin.

Aircraft were sometimes sent to Drysdale Mission, in West Australia, which had an airstrip provided with the most basic of necessities. Fuel was supplied by barge, which approached the beach and dumped the drums of petrol off the ramp. All personnel there, including aircrew, then had to swim out and manhandle the drums ashore, up onto the back of a truck, and finally pump the petrol into the aircraft tanks. Then the operation had to be flown. A large hill was at the eastern end of the Drysdale Mission strip and when landing from that direction, the Beaufighters had to drop in over it. Taking off into the hill — with a full load of fuel and ammunition — was worse.

To add to the problems, the Japanese would sometimes bomb Drysdale but they aimed for the mission buildings on the assumption that the RAAF would be occupying them. In reality, the RAAF was in tents some distance away. Sadly, bombs killed and injured a number of aboriginals at the mission.

Crews detached to Drysdale were not allowed to take mosquito nets, in case they were shot down and the Japanese would be able to deduce that they came from a forward base, not Darwin. The presence of nets was thought to indicate that the crews came from a temporary forward airstrip rather than permanent accommodation in which nets were left. Rations were scarce, as the Drysdale staff had barely enough for themselves.

On 26 August, Squadron Leader Butch Gordon led seven Beaufighters to attack floatplanes of the Japanese Imperial Navy's *934 Kokkutai* at Taberfane. These floatplane fighters had been quite active since their unit formed in March 1943, having engaged Australian Beaufighters, Dutch 18 Squadron B25 Mitchell twin-engined bombers and US 319 Bomb Squadron four-engined B24s. The Japanese had claimed about nine victories, including Beaufighters and a Spitfire in previous combats. 31 Squadron had attacked this target on 11, 17 and 21 August, claiming two floatplanes destroyed, but on 21 August Beaufighters A19-47 and -63 had collided when attacking a Japanese aircraft and both had crashed. One Japanese pilot claimed four Beaufighters destroyed in the combat.

To reach the target, the Australian force had to stage through the eastern transit airstrip at Millingimbi, departing there at 04.20 for the two hour flight to Taberfane. With fuel for 15 minutes over the target, the crews were faced with another two hour return flight, all across open sea. As usual when about 90 miles (150km) from the target the Beaufighters went down to 50 feet, pulling up when they reached it. The formation navigated via Cape Ngabordamloe and Cape Tafermaar, and as they arrived at Taberfane the element giving top cover went on to 2,000 feet while the strafers levelled off at 700, looking for targets. Many other attacks had been made with some success against the Japanese floatplanes usually moored there. The Japanese had installed anti-aircraft guns along the beach, mainly light calibre, and these opened up as the dark-green Beaufighters sped past.

Five Japanese single-engined floatplanes — three 'Rufes' and two 'Petes' — were airborne, and another Rufe was being pushed down a slipway as the strike force appeared. 'Rufe' was the Nakajima A6M2-N floatplane version of the Mitsubishi A6M 'Zero', armed with two 20mm cannon and two 7.92mm machineguns, and a top speed of 235 knots. 'Pete' was the Mitsubishi F1M2 biplane floatplane, with a top speed of 200 knots, armed with two forward-firing 7.7mm machineguns and a third in the rear cockpit. Butch Gordon fired at the floatplane on the slipway, scoring many hits. It burst into flames before his pass was completed and he was able to see that it was badly hammered by the blast of fire from the 20mm cannon and .303 machineguns.

One Pete swung in astern and attacked, but was shot at in its turn by Tom Warren, flying as No.3. Closing on the Pete, Warren fired from 100 yards (metres) in to 50 yards (metres), watching his 20mm cannon shoot pieces off the Japanese plane, which reared up then fell away out of control in a side slip only 50 feet (16m) off the water. At that moment a Rufe attacked. Warren broke away and neither he nor Doyle, his observer, actually saw the Pete hit the sea. It was claimed probably destroyed. Tom Warren believes his good relationship with Butch Gordon began with this incident. 'There were kites everywhere that day, and all our fellows had their own problems. I think Butch never forgot that.'[4]

Gordon meanwhile had gone searching along the creek banks for floatplanes which may have been pulled in there for concealment. After five he minutes could not find any and so he left. Flight Sergeant Kilpatrick, flying as No.2, had been unable to see any targets on the beach apart from the one attacked by Butch Gordon, so pulled up and fired at another Pete. He missed and the Pete turned into him. Kilpatrick fired again as they closed to 50 yards (metres). His shells burst on the Pete, which began to stream smoke from the fuselage. It whipped past them and his observer, Horton, saw it diving with black smoke coming from the engine, but it pulled out of the dive, the smoke stopped and they lost sight of it. Flight Lieutenant Taylor, No.7, watched this, and dived to attack the Pete, firing from 400 yards (metres) as they closed diagonally at an acute angle—a difficult shot—but he saw no results. His observer also fired as the Pete passed but, again, saw no result.

The formation tangled with the Rufes, and a Pete, despite problems caused by cannon jams, then broke off and left the area, with no further claims for damage to the Japanese aircraft. Butch Gordon made several other passes at the Rufes, with no result, and also left. Behind them the Rufe he had attacked first was still burning, so it was claimed as destroyed. The Japanese tried to follow for a distance, but gave up and turned away.

On 31 August, six Beaufighters led by Flight Lieutenant J.D. Entwhistle attacked again, claiming one victory, but A19-119 was lost, and it seems five Japanese fighters shared in the victory. On 17 September a return visit was paid with Wing Commander Mann leading seven Beaufighters. The instructions forbade staying to fight the Japanese but were to get in, strafe, and get out on the long trip back to Millingimbi. But as they left the Australian coast, Flying Officer Ogden, leading the top cover element of three, lost the others while flying through sea haze in the darkness before dawn. He kept on, and half an hour later met Flying Officer Ellis, one of his section. However, because of radio silence Ogden could not speak to Ellis to request him to lead. They went on in the direction of the target. Then Flying Officer White, the last of the three designated as top cover, also lost the rest of the main formation when navigation lights were turned off at dawn, as they approached Cape Lelar.

The rest of the formation approached via Cape Lelar, followed the west coast of the island, entered and searched Serwatoe Inlet, when Mann saw and attacked a large troop-carrying barge, which exploded into flames and was still burning ten minutes later. They went on to the Maikoor Channel and approached the target from the south-east.

Cloud had been thickening and cloud base was now at 2,000 feet, with 9/10 stratus.

As Mann approached the floatplane beach, he saw a Pete and a Rufe above and, as there was no top cover, he pulled up with Flying Officer Mitchell, No.2, to attack, while Butch Gordon and Flight Lieutenant Jim Gordon went in to attack as planned. Mann and the Rufe were approaching head-on; he fired but was then badly hit in the starboard wing by anti-aircraft fire so jettisoned his bombs and turned away.

Mitchell had fired at a Rufe which half-rolled and dived after Butch Gordon, apparently unhit by Mitchell's fire. The heavy volume of anti-aircraft fire coming from the beach drew Mitchell's attention, and he dived to drop his two 250-pound bombs into the trees there, but saw no results, continuing after Mann, turning along the coast to Serwatoe. Flight Lieutenant Jim Gordon had strafed and bombed the beach defences and left, watching a Pete flying some 800 yards (metres) away, but it did not attack.

Butch Gordon had pulled up with the others, then dived on a Pete, firing at it until they were right onto the coconut trees, despite being followed by a Rufe which was shooting and coming after them through the Japanese anti-aircraft fire. The Rufe pilot was not putting his nose down enough to lead the diving Beaufighter, and his tracer was flicking over the top of Ron Jordan's cupola. Butch Gordon always told him that 'the Japanese could not shoot for nuts', but Jordan's reply was that he was the one who saw it coming, not Gordon. The Pete, shaken and riddled by the cannon and machineguns, burst into flames and was claimed as destroyed, confirmed by Flight Lieutenant Gordon. At this time, Jordan saw Mann's aircraft hit by the anti-aircraft fire.

Butch passed on over the coconut trees and dropped one 250-pound bomb which hit a machinegun position in a clearing, the explosion hurling sandbags and gunners into the air, just as a 12.7mm machinegun round hit the cockpit from the front, exploded, and blew the petrol gauges off the panel. Fragments hit the parachute and Gordon and Jordan — each on the right arm. The Beaufighter had armoured doors between the pilot and observer, but Jordan preferred to keep them open, to see the pilot. The 12.7 fragments flew along the fuselage and speckled his right arm, so tiny that he described their wounds as 'like chicken pox.'

Gordon jettisoned the other bomb, as it would not drop on the normal release. At this time, a Pete attacked and the Rufe turned away, Butch turned left and avoided the Pete, searched the beach unsuccessfully for more targets, flew on to Serwatoe where the barge was still burning and then continued on to Australia. Flying Officer Mitchell joined him, and they then came on Mann, struggling back with a badly damaged Beaufighter.

Meanwhile, the other members of the formation, who were to provide top cover, had arrived in the target area, and Flying Officer White was attacked twice by Rufes as he searched the Maikoor and Cape Lelar areas, then met Ogden and Ellis and set course for Millingimbi.

On Mann's Beaufighter, the wing skinning outboard of the fuel tank had been stripped away and the aircraft was barely controllable. On several occasions it almost went into the sea, but each time Mann kept it up — somehow — out of the water. The Beaufighters were fitted with Hercules MkII engines, and were very fast at sea level. However, their range was only 1,000 miles (1600km), which did not allow much extra when flying the long over-water missions, and the greatest shortcoming was the lack of feathering propellors. A wind-milling or stationary propellor literally became a drag on the aircraft, which required petrol use to overcome, and could result in the aircraft failing to return to the Australian mainland. Mann was in a real predicament.

Ron Jordan sent messages to Darwin each time it seemed

Mann's Beaufighter would hit the waves, asking for a direction finder bearing on the position, but Mann was able to reach the coast on one engine, which then stopped, and he put the heavy aircraft down into one metre of water in a swamp, about 15 miles (24km) south-west of Millingimbi. Mann and Flight Sergeant Harber climbed out whilst Gordon circled until a DF could be taken of the location and rescue operations commenced.

Elsewhere, Ogden and Ellis had been forced to land on distant beaches when they ran out of fuel, but replenishments were flown in to them by a Hudson from Darwin. Ellis' problems were compounded by a bird-strike when landing, resulting in damage to the oil cooler and oil pipes. Ellis crashed on take-off and A19-30 was reduced to component parts.

The mission had cost the RAAF more than it had the Japanese, but this was part of the cost to be expected when operating at extreme range in the conditions prevailing at the time. Japanese expectations were greater, and their pilots claimed six Beaufighters shot down or made to force-land due to damage.

On 9 October, Gordon led six Beaufighters on what was described as an 'harrassing attack' on Selaroe Island, to be followed by armed reconaissance around Tanimbar Island, north of Darwin. Designated targets were allocated to specific aircraft, and if no shipping was found the remainder of the bombs carried were to be dropped north of the jetty onto Larat. The order stipulated that natives fishing from canoes were not to be attacked. Butch Gordon and Ron Jordan, in A19-40, were leading the other five through 5/10th cumulus, base at 1,000 feet, visibility ten miles (16km). Flying time to the target was two hours, as was the return. The flight out would be at 1,000 feet, until 50 miles (90km) out, when they would descend to sea-level.

The formation made landfall about one mile (1.6km) south

of Kerval Creek, and turned for the target, some bombing and others strafing the suspected stores dump areas. They roared on to Werain, Gordon strafing a whaler en route, and the formation flew on to rendezvous at the north-east end of the island.

Pilot Officer McCord, in A19-148, was last in the formation and saw a Sally bomber coming in to land. He ignored some slight, light and inaccurate anti-aircraft fire coming from the middle of the strip area, swung the big Beaufighter through a 180-degree turn, dropped his bombs on the strip itself, and went for the Japanese bomber. Its pilot had seen him, retracted his wheels, and was climbing for the cloud-base at 2,000 feet. McCord was at the 8 o'clock position, half left-rear, and the Sally pilot, realising he could not out-climb the Beaufighter, tried to turn into him. McCord swung around behind the bomber, turning with it, and opened fire at 400

Japanese Mitsubishi 'Sally' bomber hitting the sea off Selaroe, 9 October 1943. (Ron Jordan)

yards (400m), closing to 50 yards (50m). The hail of 20mm cannon set fire to the Sally's starboard engine, then the port. The starboard wing and engine fell off and the Sally rolled into its final dive just off the end of the strip, 100 yards (100m) into the sea. One person was seen to try to parachute from the Sally, but his parachute was on fire.

Flight Sergeant Childs had been photographing the stricken Sally's last moments. Then McCord found that his guns and intercomm were useless, so he called Gordon and told him he was returning to base. He had fired 455 rounds of cannon. Only one gun had worked properly — one having a jammed belt and two others feeding slowly. No machineguns were fitted to the Beaufighter. The tail marking of the Sally was later identifed as that of the Surveyor General of the Japanese Army, and it was believed that he or senior staff were aboard.

Meanwhile, Flight Sergeant Ferguson had gone to bomb Adaoet and the other four Beaufighters were near Namtaboeng. Ferguson bombed and saw a large group of people, probably Japanese, standing outside a big hut. He strafed them, believing the concentration of machinegun and cannon killed about 20. Then a twin-engined Japanese aircraft attacked. At first it was not identified, but later recognised as a Nick, the Kawasaki Ki45, with a top speed of 340 mph (540kmph) and armed with 20mm cannon, 12.7mm machineguns and a 7.92mm machinegun for rear protection. The Nick came in from behind the RAAF fighters, attacking Gordon at a height of 800 feet, firing from 300 yards (metres) away, and passing him at the 2 o'clock position, slightly to the right front. Gordon turned after it as the Nick flew away south. The Japanese passed over the top of Muggleton, in the fifth Beaufighter, and he also turned after it. Gordon dropped his bombs and accelerated to 250 knots, closed to point-blank range of 25 yards (metres) near Kerval Creek and opened fire, ignoring the slight weaving of the Japanese pilot who was trying to allow his reargunner to shoot down the oncoming Beaufighter. The gunner did hit A19-40, holing the upper mainplane and shooting off the starboard exhaust ring. Part of the aileron was shot away, and one bullet passed between the two wires controlling the aileron itself. But Gordon's 20mm cannon and .303 machineguns hammered home. The Nick jerked up into a stalled left-hand turn and Butch called to Jordan to look at the effect of the gunfire. Jordan did so and saw it 'just flying to pieces'. Parts flew off the Nick, it tilted upward and then crashed down into the sea.

As it was going down, Garnham and Stubbs also put bursts into it for good measure and Muggleton, coming along fourth, watched as it crashed. Gordon checked his aircraft, assessed the damage and continued on to strafe a powered barge near Larat. Coming back over Selaroe, he told Flight Sergeant Ferguson to take over as formation leader whilst

Damaged starboard aileron on Beaufighter A19-40, 9 October 1943. (Ron Jordan)

Gordon dropped back to 1,000 yards (metres) astern of the others.

Three minutes after leaving the island, Jordan called that they were being attacked from behind. A Nick had jumped them out of the cloud cover and was 200 yards (metres) away, firing and hitting them. At least six rounds hit Jordan's machinegun, ripping it out of his hand and tearing his flesh whilst other rounds were stopped by the ball-bearing mount of the gun. Other bullets hit the flare floats stacked inside the fuselage near Ron. More hits flashed on port engine and starboard aileron, smoke poured out of the port engine and they dived towards the sea, with Jordan calling on the radio, 'Shit, we're hit', which bellowed out of the loudspeakers back at Coomalie. He quickly asked for a DF to mark the spot, then the Nick was past, sliding out to take up a formating position some 400 yards (metres) to starboard.

Destruction of first Ki45 'Nick' by Butch Gordon, 9 October 1943. The aircraft has gone into the water. (Ron Jordan)

The Japanese flew there for a moment, probably assessing the situation, then made a decision. The Beaufighter was obviously badly hit and going down; he surged the Nick forward, crossed ahead of Gordon and moved onto the tails of the Beaufighters in front. He opened fire on Ferguson at about 1,000 yards (metres) range but Ferguson's observer,

Kawasaki Ki45 'Nick' in flames, shot down by Butch Gordon, 31 Squadron RAAF, 9 October 1943. (Ron Jordan)

Blades, fired back. The Nick swung away to port whilst Ferguson opened his throttles and pulled further ahead. In those seconds, Gordon opened up both engines, closed up on the Japanese and fired a deflection shot from 350 yards (metres) behind. Fire came from the port wing root of the Nick. It began slowly losing height and then went into the sea about 15 miles (24km) off Selaroe.

At this time, service in the Darwin zone was not recognised as war service and pays were fully taxed. Gordon would

Beaufighter A19-40 after landing at Livingstone strip, 9 October 1943, after destroying two Ki45 'Nicks'. (Ron Jordan)

comment, after each shooting-down of an enemy, 'Take that off my tax, you bastards!'.

Now Gordon was two hours from home, with a long overwater flight ahead and one engine damaged. He throttled back the port engine, then switched it off and continued with the starboard, giving it maximum boost to maintain height. The aircraft was very hard to control, and they were at 800 feet at 105 knots, sometimes only indicating 80 knots, at which speed the Beaufighter was not supposed to be able to fly. Ron Jordan sent a regular stream of messages requesting bearings to fly, and to allow DFs to be made of their position. He was wondering how Butch Gordon, like Mann on the other mission, managed to keep the big aircraft flying.

As he was passing over Bathurst Island, Gordon checked his hydraulics, but a broken airline had made them unusable. Preparations were made to receive the Beaufighter at Coomalie, but fuel was running out, and Gordon crash-landed on Livingstone Strip, swinging the Beaufighter off the runway as much as possible. He subsequently apologised to the Spitfire squadron CO for leaving the tail of the big fighter poking onto the strip. There were 50 cannon and 120 7.7mm machine-gun holes in Beaufighter A19-40. The aircraft was written off.

Garnham also had an adventurous return, running out of fuel and landing in a minefield just to the north of the RAAF airfield at Darwin. Luckily no mines exploded. The Army unit which laid them then had to do some re-thinking.

No shipping had been seen and though it was obvious some work was being done on the airstrips they passed and the roads appeared to be in use, all the villages seemed to 'be dead', as the crews reported. For a few weeks in October the squadron moved to Darwin, and were housed in the old, bombed RAAF buildings there. The more established units made no use of these places but to the squadron coming from tents and huts at Coomalie, 'it was like Christmas.'

On 2 November 1943, a mission was flown to attack the new Japanese air strip built at Trangan on Aru Island. Three Rufes intercepted and as one came in from behind, Ron Jordan fired and hit it with his rear gun. He watched it begin to smoke and go down but in the combat it could not be followed, and no one saw it crash.

On another sortie, on 11 November, Gordon told the crews to come up line abreast in the attack run, to divide the amount of anti-aircraft fire and make it harder for the Japanese. However, an expert Japanese gunner nick-named 'Sumalaki Sam' was on duty that day, as they attacked the airstrip, and his fire forced some of the wingmen to break away, leaving Butch Gordon and Ron Jordan to face the fire alone. Butch went right down onto the trees, firing all armament, and the cannon became so hot it exploded, forcing out the bottom of the cannon bay in the fuselage belly. Ron could see the water streaming past below them, then the airstrip, and flashing glimpses of Japanese down there as the Beaufighter hurtled low overhead. Ron has a special reason for remembering this particular occurrence as it took place on his birthday. On landing they were told of the award of the DFC to Butch, for the sortie of 9 October.

On 21 November 1943, six B25s from 18 Squadron, the Dutch bomber squadron, were to go on a shipping hunt in the Maikoor area with six Beaufighters to provide cover. The crews were told that the ships were carrying guns and radar to both improve the Taberfane defences and for installation at the new strip at Trangan. The highly-trained Japanese technical personnel were as important, if not more so, than their equipment. Their loss from death or wounds would be a severe blow to the enemy. Again it was a long flight over water to and from the target and on this occasion conditions were fine, with visibility estimated to be 20 to 30 miles (30–50km). The Beaufighters would approach at 1,500 feet, go down to sea-level for the last 100 miles (160km), climb to

The beached remains of a Japanese escort vessel after a shipping strike by RAAF Beaufighters and Dutch Mitchells. (RAAF official)

2,000 feet in the target area and return to base at 6,000 feet. Butch Gordon led, with Ron Jordan as observer, and the other crews were Gaunt/Jones, Archer/Holland, Gordon/Spencer, Ferguson/Blades and Gerdes/McMillan.

The formation arrived over Hokma village then turned left to Taberfane, the Beaufighters climbing to 2,500 feet to cover the B25s which were making mast-height attacks on ships in the bay. There was one freighter of about 2,500 tons and two smaller 500-tonners, with many small craft going to and fro unloading the cargoes. Intense flak from the shore and the ships came up and this, plus smoke from bomb bursts made, it impossible for the Beaufighter crews to accurately assess the results of the B25 attacks. B25 N-159 was hit by anti-aircraft fire and crashed at the mouth of the bay. Sergeant de Putter and crew were all killed. Hits and near misses sank the freighter and damaged the other two ships.

Then two Rufe floatplane fighters were seen taking off. Butch Gordon ordered the second flight to attack the Rufes and Gordon, Ferguson and Gerdes dived. Jim Gordon made one pass, firing at the leading Japanese, who turned under the Beaufighter, avoiding the fire. Gordon climbed back to 2,000 feet, waiting for the B25s to cross the target. Behind him, Ferguson had been unable to fire, so had climbed out of the combat to look around. Gerdes had dived, firing at one Rufe which took off through the splashes flung up by the Beaufighter guns as Gerdes came in from behind. Gerdes whipped around and flew in behind the other Rufe, firing from 250 yards (metres) in to 150 but the Japanese turned sharply right and evaded. Then both Rufes went for Gerdes, closing on his tail, firing and breaking away when McMillan shot back with the .303 machinegun.

Ferguson called to Butch Gordon that Gerdes was in trouble and Butch made a head-on attack on one of the Rufes, which held course until 100 yards (metres) away and then rolled onto its back and tried to dive away. Gordon followed in an aileron turn but the floatplane burst into flames and went on down into the sea. Ron Jordan, in the back seat, was taking in all of this through his cupola. The large ship was sinking, leaking oil, after the B25s had passed. To add to the fray, a third Rufe, probably from nearby Serwatoe, sailed into the combats.

A Rufe had fastened onto Ferguson, who had gone up to 2,500 feet, and he increased power to pull away from the Japanese who broke off and turned while Ferguson climbed to 4,000 feet. Blades called to him that there was a Rufe below. Ferguson dived, fired and missed, and zoomed back to 3,000 feet. Blades called there was a Rufe on the tail of a Beaufighter, Ferguson dived head-on, firing to 100 yards (metres). Both broke, the Rufe began climbing, with smoke coming from the engine. Ferguson hauled around again, and the Rufe was going down with Blades shooting as it passed below.

All this happened in such a short time that as Gordon's destroyed Rufe hit the sea 500 yards (metres) from shore, it was followed almost at once by Ferguson's, which fell in water so shallow the wreckage continued to burn.

Meanwhile, Archer had attacked one Rufe shooting at a Beaufighter, and closed to 75 yards (metres) when the Japanese flick rolled and dived, evading Archer, who went down to sea-level and circled the area. He saw no one, so set course after the B25s. In the melee, he had a wingtip and the observer's cupola shot away. Holland was lucky; the shots carried away his headphones, without injuring him. Gerdes had been hit by flak, which blew a foot wide (30cm) hole in the port wing, peppering the fuselage, punching holes in McMillan's cupola, ripping into his parachute, puncturing the starboard fuel tanks twice, severing the airlines so the guns would not fire. McMillan was also lucky. He suffered no injury although a pencil was shot out of his hand and a wrist identity disc was shot away.

The four surviving B25s were going south, covered by Gordon, and the Beaufighters gathered to go with them. Ferguson saw a fourth Rufe taking off from Serwatoe but it headed off south and made no attempt to fight. By the time they had cleared the area, the largest ship had sunk completely, leaving only an oil patch on the waters.

The Beaufighters landed at Millingimbi to refuel, then set off back to base at Coomalie. In one of aviation's mysteries, Gaunt and Jones, in A19-145, disappeared between Millingimbi and Coomalie. No trace of them was ever found.

When the damage and action had been tallied, some found it hard to believe that people could survive the flying metal in and around the Beaufighters.

On 16 December 1943, Wing Commander Mann led eight Beaufighters to attack two merchant ships and an escort vessel reported off the coast of Timor. The ships were expected to be around Lautem but if they could not be located, Dili was to be attacked. Again it would be a four-hour sortie, at sea-level for the last 100 miles (160km).

Only six aircraft actually flew the operation, as Ellis could not take off due to fuel feed failure and Wickens returned shortly after take-off with engine trouble. The formation arrived at the coast, then flew over the mountains, passing a Japanese radar station on a prominent hill, and steered for a column of smoke rising from a ship bombed by the Dutch B25s the day before when they had hit two ships, leaving them on fire.

Mann lead the formation over the ridgelines, down into the valley of the Malai Ladar River, and the Beaufighters streamed down to the sea, below the hill crests, emerging into Lautem roadstead to see the third ship, the 3,200 tonn *Genmei Maru* unharmed, with two barges alongside and three others between ship and shore. Surprise had been achieved.

The anti-aircraft guns in the commanding position on the cliff tops were unable to fire at the approaching Beaufighters and even when they had passed the aircraft were so low that the guns could not depress far enough. The aircraft were travelling so fast that the Japanese gunners seemed to have problems traversing and changing the range settings.

Butch Gordon, who did not carry bombs, fired his cannon and machineguns at a barge full of troops which was alongside the ship, setting it on fire. He then swept up to act as top cover while the others raced in to bomb. Ten 250-pound bombs were delivered onto the ship and barges. Mann put his two at the base of the foremast, Ferguson and Archer each hit the ship amidships and Quance hit it at the stern. Eleven seconds later the explosions blew up decking and superstructure, fires broke out at once and a great cloud of smoke uncoiled with debris flung up by an explosion from Quance's hits. Garnham made two direct hits on the centre of the three barges midway to shore and the centre barge capsized with a total loss of cargo. Many troops were on these barges and the number killed or wounded would have been considerable. Ferguson was last in and smoke from previous explosions was so thick that he almost went too low, only just clearing the ship by flying between the masts, carrying away the radio aerials.

Ferguson then climbed to assist Butch Gordon who was engaging two Nicks which had been patrolling the area. They were all three circling at 2,000 feet and, as Ferguson went for them, Smith—from his back seat—saw two more Japanese fighters coming from Fuiloro. One Nick was making a stall turn as Gordon attacked, firing to within 50 yards (metres), and he noticed the Japanese chocolate and bright green camouflage, red hinomaru on the wings and open gunner's position in the rear of the cabin. Then the Kawasaki fighter went down into the water, breaking in two on impact.

Archer had been fired at inaccurately by a gun at the mouth of the Malai Ladar River. He gave it a short burst in return but saw no results. Garnham, who hit the centre barge, was set upon by a Nick, who managed to put one 12.7mm bullet through his tail while Garnham flew over Fuiloro strip and again inaccurate ground fire was aimed at him. He swung back to the remainder of the formation. Meanwhile, over the target, Gordon made a head-on attack on another Nick, firing at 200 yards (metres), hit it in the wings with .303 inch (7.7mm) machineguns. Gordon saw pieces fly off but the Nick kept on going. Below, the attacking formation was leaving and Gordon went to join them. The anti-aircraft fire was becoming more accurate, and explosions followed them for some time, inflicting a shrapnel hit in the port engine nacelle of the Beaufighter.

One Nick followed the Beaufighters for 80 miles (110km). The Beaufighters found that at 210 knots the Nick could overtake; at 220 he could maintain position; at 230 could not keep up and at 240 knots at sea-level the RAAF fighters could easily draw away. Wing Commander Mann and three other Beaufighters were in line abreast and Mann varied speed several times to check the performance of the Japanese. His radio failed and he could not organise an attack on the Nick, the pilot of which presumably thought he had four frightened aircrews ahead. Finally the Nick turned back.

Again cannon stoppages and problems affected some of the aircraft. Mann's worked well but Gordon fired a total of 225 rounds from his four cannon with one slow-feeding and another jammed by a case caught in the feed way. Ferguson also had one cannon which lost tension in the feed apparatus.

On 4 January 1944, Butch Gordon led eight Beaufighters as escort for six Dutch 18 Squadron B25s attacking shipping in the Tenau-Koepang area. Cloud base en route was 1,500 feet with 4/10th broken cumulus but in the target area there were heavy rain squalls with visibility much reduced.

As the formation was approaching Cape Mali, Flight

Lieutenant Jim Gordon reported a Betty bomber approaching from 2 o'clock, 1,200 feet above. This was the Japanese Navy's Mitsubishi G4M twin-engined bomber. When the Betty passed behind the formation, Butch Gordon called Archer, flying as No.2, to join him and he turned right after it, swinging behind the Japanese. Gordon fired from below, coming up level, while from his rear cupola Ron Jordan watched the Japanese return fire from the side guns and with the 20mm cannon in the tail. Then the Betty began a right-hand spiral dive. Butch Gordon was firing from above, hitting it, setting both engines on fire, with flames coming from both wing roots. They pulled up alongside and could see Japanese moving around inside. The Betty began to go down. The starboard wing flew off, followed by other pieces, and the blazing remains hit the sea and exploded. No one bailed out. Why the pilot took no evasive action, but continued to fly peacefully across the formation, will never be known. Perhaps he never saw them or, if he did, perhaps he hoped that if he did nothing they would not notice him.

Butch Gordon turned back and slid into place in the formation as before. As the Mitchells neared Semau, a flight of single engined fighters was seen, which meant they had to be Japanese as there were no single engined Allied fighters able to reach the area from Australia. Butch Gordon took the Beaufighters out to sea to attract the fighters and give the bombers a clear run. But apparently the fighters did not see them. Through heavy rain, a stationary ship of about 4,000 tons was seen near Tenau and the B25s attacked. Gordon strafed a barge which blew up and before he could avoid them, pieces of debris hit and penetrated the leading edge of his wing. Archer, Mitchell and Rinkin attacked a local boat, all hitting it, shooting the mast off but no fire resulted. The Beaufighters picked up the B25s as they emerged from the rain squalls and all set course for base. To avoid interception

Gordon and Flight Sergeant Rinkin climbed 5,000 feet above and ten miles away from the main formation, but nothing eventuated.

It was decided that a second strike of three Dutch B25s was to attack shipping in the Tenau-Koepang area, with four Beaufighters to provide top cover. Gordon, with Spencer as observer, led Wickens/Staines, Ferguson/Blades and Quance/Taylor. As normal, the last 100 miles (160km) would be at low level, but on this day they would fly through a front some 200 miles (320km) from the target with visibility, when not in rain, of some ten miles (16km). After take-off, Gordon, Ferguson and Quance set course for Port Paterson, thinking they were following the B25s. On arrival at Por Paterson, no B25s could be seen, so they circled. No bombers arrived, so they went to the target area but again found no one, so returned to Coomalie.

Meanwhile, Wickens had found the B25s and went with them to Tenau. They attacked a 4,000-tonner, making direct hits and leaving it burning. As there were no enemy aircraft in sight, Wickens strafed a 45-foot local boat filled with personnel. Fire from the ship hit N5-137, flown by Captain Rees, in the starboard wing and the B25 swung toward Captain Holswilder, who had to take sharp avoiding action. Holswilder then continued his attack, scoring a hit but Rees struck the ship's mast, going out of control into the sea only 75 yards (metres) on the far side of the ship. Wickens formated on the two remaining B25s and they returned to base.

By 27 February 1944, due to his recent period of intensive flying, Butch Gordon had his flying activity greatly reduced by the squadron commander. On that day, Ron Jordan checked with Gordon, who said there was no flying intended so Ron agreed to take a swimming party from the squadron area. Before they left, he heard that Butch was to fly, but could not find him to confirm this, so went swimming anyway.

Several Beaufighters had been having problems with propellors and after work on A19-165, Gordon and Flight Sergeant Smith took it up to test it at 17.21 hours. This aircraft was the squadron's first with feathering propellors. Watchers saw them circle the strip twice at low altitude then, with the starboard propellor fully feathered and the port windmilling, it passed behind a range of hills at only 200 feet. Both engines failed in succession. Gordon attempted a belly landing but was confronted with timber ahead and outcrops of rock among the trees. The Beaufighter hit and immediately burst into flames. Smith was killed inside the fuselage but Gordon was thrown out, or managed to climb out, and was found some 50 yards (metres) from the wreck, suffering from extreme burns and shock. Within 30 minutes, he was admitted to hospital but he died two and a half hours later. Ron Jordan, with the swimming party, telephoned the squadron to check if things were alright, as some of the swimmers wanted to go on to the open-air film. He was told of the crash and of Gordon's injuries.

By his spirit, skill and leadership, Butch Gordon had played a large part in sustaining the aggressiveness and morale in 31 Squadron during its operations in an inhospitable environment, at the limits of the range of its aircraft and against an enemy who was skilful and aggressive. On the relatively rare occasions when air combat was possible, and when he was in a position to engage, Butch Gordon was able to destroy an enemy aircraft. He was the squadron's leading pilot in gaining air-to-air victories in the short time he was on operations in the Darwin area.

Appendix

Squadron Leader Reginald Gordon DFC*

Reginald 'Butch' Gordon had 2,900 hours total flying time, and had been graded 'Above Average' at OTU and in 31 Squadron. He had flown 25 strike operations, 12 convoy escorts, five other sorties and two navigational training flights, for a total of 42 flights with 31 Squadron. His victory claims are:

Date	Beaufighter	Aircraft	Location
26 August 1943	A19-70	one Rufe	Taberfane (ground)
17 September 1943	A19-40 EH-G	one Pete	Taberfane (ground)
9 October 1943	A19-40 EH-G	two Nicks	Selaru area
21 November 1943	A19-140	one Rufe	Maikoor
16 December 1943	A19-144	one Nick	Lautem
4 January 1944	A19-149	one Betty	Cape Mali

We never met his equal

Intruder Mosquito

Sqn Ldr Charles Scherf DSO DFC*

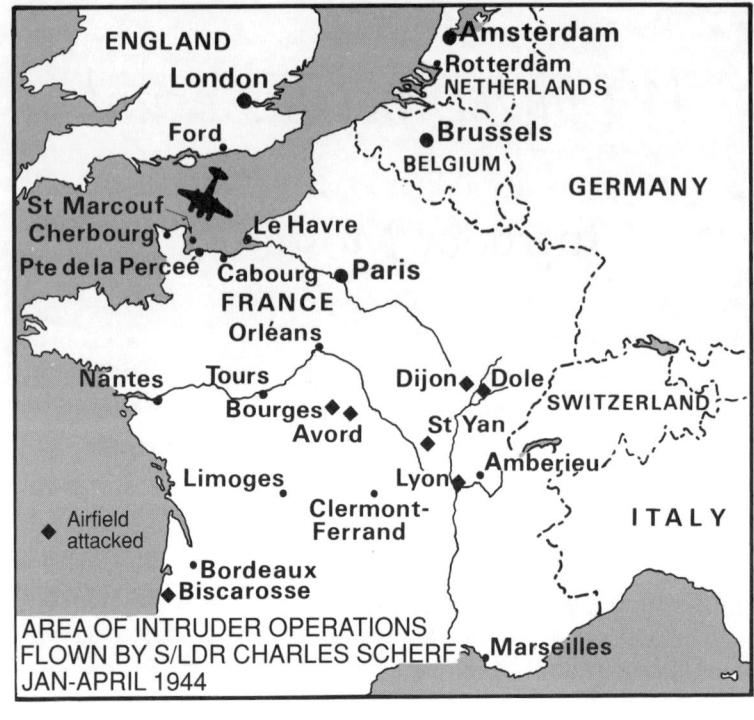

Charles Scherf DSO DFC

Charles Scherf's combat career got off to a slow start, then suddenly seemed to accelarate, with each successive sortie more spectacular than the previous mission, until he himself decided to call a halt. In flights deep into German-held Europe, he destroyed 13.5 enemy aircraft in the air, ten on the ground and damaged a further seven and was awarded a DSO, DFC and Bar.

Charles Scherf was born on 17 May 1917 at the Vegetable Creek Hospital (later Emmaville Hospital) which, at that time, did not have a maternity section; local authorities and senior doctors did not believing such a facility was necessary given that their mothers had not required it. Charles was the largest baby recorded as born in the state of New South Wales at that time, weighing 13 pounds (6kg).

His parents were quite active in community affairs, and owned 'Big Ben', a sheep grazing property. The home surrounds included tennis courts, a nine-hole golf course and river frontage for swimming. Ponies and horses were available for children and adults. The children had to feed the chickens, look after their own horses and do other chores including rake a large gravelled area in front of the house.

Charles and his sister Beryl attended a small country school which comprised only five pupils from two families. The sole teacher lived on a month-about rotation with these families during school terms. No doubt this meant close attention to homework and the availability of extra tuition. The school itself consisted of a one-room building with a rain-water tank alongside, a decrepit tennis court, lavatories some 300 yards (metres) away in the trees and a couple of horses tethered nearby. Later, in the UK, when Charles was asked if he went to a good school, he could say that he had gone to a public school but he did not explain that this meant the opposite to the British 'public school'. The family took amusement from the embellishments to the local school which Charles gave to the British. The school was five miles (9km) from 'Big Ben' and from the age of eight Charles rode his pony, summer and winter, to and fro. At one time he also played hooky, going tin-mining instead of to classes. As usual, he was soon found out.

The Scherf grandparents were German, Mr Scherf being an engineer who came to Australia in the 1854 goldrush to make his fortune but taking up the property instead. All their children and grandchildren were given tuition in musical instruments and in singing. The children of both generations went to sleep to the strains of the great classicists, and the younger also liked lighter music and jazz.

From an early age, Charles displayed his native courage. At the age of ten, near Emmaville and in the horse-drawn sulky with his mother, the animal bolted. Charles pushed his mother down to a safer position, grabbed the reins and held on, hauling back and somehow managing to halt the horse in the streets of the small town. At 17, whilst at the coast, he rescued a man from wild surf which onlookers did not dare attempt due to the wild conditions. Scherf could not abide cruelty to animals. On one occasion a travelling salesman was seen mercilessly flogging his horse, forcing it to haul the salesman's wagon from the mud at the creek crossing close to the 'Big Ben' homestead. Charles called to his sister Beryl, ran to the stable and grabbed a whip, jumped on his horse and rode to the scene, delivering a whipping to the man, telling him, 'Now you know how the horse feels!' The adult Scherfs had come out onto the verandah to see what all the excitement was about, and the salesman curbed any desire to retaliate.

By the time he was 17, Charles Scherf was a well-known local footballer and member of the local (Army Reserve) Light Horse unit, winning many prizes. His sister Beryl remembers him as a 'calm child, loyal, and it was hard to visualise the man who evolved—stern faced, lean, with great strength of character.' She concedes that he would also have been a bit of a wild youngster in his football and Light Horse days. Keenness for sport interfered with Charles' musical practice, though he could have developed into a skilled violinist and pianist. He was talented enough to imitate the big names of the day, such as Jimmy Durante, Carmen Cavallero and Jimmy Dorsey.

At the age of 19, Beryl married an American and went to live in San Francisco.

Charles took 5,000 sheep on agistment to the O'Hara property, where he met and fell in love with Hope O'Hara and they were subsequently married. When war came in September 1939, he wanted to go but was needed on 'Big Ben'. Finally, he managed to enlist. None of his family imagined he would become a famous fighter pilot.

Charles Scherf joined the RAAF in September 1941 and was a member of No. 20 Empire Air Training Scheme (EATS) Course, undertaking Elementary Flying Training (EFT) at Temora. While there, even though only a new member of the RAAF, his personal qualities as a man and a friend were noticed, and recalled, by L. Morrison, another of the trainees. They were in the same hut, and became good friends.

'Charlie was a natural,' said Morrison. 'While I was still trying to get my landings right Charlie was looping and rolling the Tigers with no trouble at all. In the early stages I doubted if I was cut out to be a pilot but Charlie helped me to become more confident and insisted that I would become a good pilot. His confidence rubbed off on me, and I survived Temora.'

Both Scherf and Morrison graduated in February 1942, Scherf assessed as 'Average' with the comment 'Inattentive to

Detail', but by the time he graduated from 6 SFTS at Mallala, South Australia he had improved to 'Above Average' and the comment was 'Sound Pupil Pilot.' It was June 1942, and he stayed at Mallala for Advanced Training which continued until 16 September. By this time he had 218 hours, and was an 'Above Average Pilot' but an 'Average Navigator'.

Scherf went to the United Kingdom for further training, meeting Morrison again when they sailed to England on the same ship. Many of the draft wanted to become fighter pilots, but by then the RAF was on the offensive and bomber pilots were needed. However, Charles Scherf managed to avoid this, and went to fly the famous De Havilland Mosquito twin-engined fighter. He graduated from 60 Operational Training Unit (OTU) on 5 July 1943 with 382 hours and was a 'Good Average Pilot', 'Average Navigator' and achieved a 'Good Average' in air gunnery. The comment now was 'Apt to be careless at times'. Scherf's training had taken almost two years and he was now deemed ready for operational flying.

He was posted to 418 'City of Edmonton' Squadron RCAF, commanded by Wing Commander Paul Davoud DFC. Scherf began training to reach the high standard necessary to take part in operations from RAF station Ford. For the remainder of July Flying Officer Scherf and Flying Officer Brown, in the squadron's B Flight, did cross country flights, night-flying tests, air-to-air firing practices and single-engine flying.

When the German night bombing offensive over England began to lose intensity, the RAF was able to increase its own night offensive operations, called 'intruding', over Europe. Suitable aircraft were not available at once though Bostons, Beaufighters and Hurricanes were used at first. Later the superb Mosquito was used in this role and eventually became the prime intruder aircraft. Great problems of suitable aircraft, control of them and training for offensive operations had to be overcome and some time was to pass before the intruder effort achieved much weight. In January 1943, only 604 Squadron RAF and 418 Squadron RCAF were full-time intruder squadrons, equipped with Bostons, but with Mosquitos soon to arrive. The Intruder Controller at HQ Fighter Command issued nightly operational orders, but the squadron commander had some leeway in deciding height and route for

A 418 Squadron RCAF 'City of Edmonton' Squadron Mosquito coming in to land. (Mary Stewart)

the crews in his squadron. By the end of 1943, incomplete Luftwaffe records show a loss in France of 30 aircraft, possibly to intruders, and at least 18 in Germany. Scherf arrived in 418 Squadron as the intruder operations were developing effective techniques to take the war with the Luftwaffe to the limits of the range of the Mosquito. The squadron had begun re-equipping with Mosquitos in May 1943, phasing out the Douglas Boston. The final Boston sortie was on 8/9 July, the squadron's 532nd on that type. With it, 418 had destroyed 5½ enemy aircraft and damaged six.[1]

On 24 July 1943, the RAF launched a devastating series of attacks on the German city of Hamburg, during which

'Window' tin-foil strips were used to confuse the German radars. The defences were overwhelmed and the circumstances combined to create the first man-made firestorm which destroyed much of the city and killed an estimated 40,000 people. Morrison, Scherf's friend from training days in Temora and the voyage to the UK, was now a Lancaster bomber pilot on 408 Squadron RCAF, piloting a crew composed of himself as the only Aussie, four Englishmen and two Canadians. Their tour of operations began with these Hamburg raids.

In August Charles Scherf began flying intruder missions over France, to Criel, St Andeux, Orleans, Le Crotoy, Javincourt, Dreux, Chartres, Ault, Florennes, Vendeville, Vitry, Laon and other places, sometimes bombing and strafing trains or other targets. Apart from close anti-aircraft fire, he found little to note in his log book. In September, he flew to Germany then, with two other squadron crews, began special training with Lancasters of the famous 617 Squadron, the 'Dambusters', and on the 15th flew as escort when they dropped 12,000-pound bombs on the Dortmund-Ems Canal.

In the Mediterranean, the Allies had occupied Sicily and invaded Italy; Mussolini was held prisoner but was rescued by Otto Skorzeny; Eisenhower and Marshal Badoglio announced the surrender of Italy but the Germans there, with Italian Fascists, continued fighting. The Russians were advancing again and the Japanese were being forced back in the Pacific by US and Australian operations. On 18 November, the RAF made its heaviest raid on Germany, dropping 350 4,000-pound (2,000kg) bombs on Berlin in half-an-hour. This was part of 'Bomber' Harris' onslaught on the capital of the Reich — a series of 16 heavy raids totalling over 9,100 sorties — which would cost 492 four-engined bombers lost in action, another 95 destroyed in crashes and a further 859 suffering serious damage, with some 3,500 aircrew lost.

For Charles Scherf, more intruder operations continued over France, Holland and Germany, with the occasional train being attacked. The weather became worse, and was mentioned more frequently in reports. On 25 November, Scherf and Brown went to the Bonn area but 'experienced icing at 4,000 feet and St Elmo's Fire on propellor and wingtips. Target not patrolled because of 10/10 cloud at 2,700 feet.'[2]

Scherf described the realities of life on an operational squadron to his family, as far as the censor would allow — one example being the way in which the eggs of those who did not return were eaten by the survivors. From San Francisco, his sister Beryl regularly sent parcels containing blue shirts, socks, soap, chocolate and US scotch. Scherf particularly liked the smell of the cakes of Cashmere Bouquet soap, telling Beryl that it was fresh and 'great after the smell of showers, damp clothes and the male smell in the barracks.'

Then on 28 November it was decided to take the Mosquitos across in daylight. Flying Officer Jim Johnson, from Ontario, had trained with Scherf at OTU, and planned the daylight Ranger with his navigator and the Intelligence Officer, Hopgood. The crew originally included in Johnson's plan as the second Mosquito crew were on leave, so Scherf joined him. Johnson, his navigator Jim Gibbons, Scherf and Brown set off at 15.45 for the seaplane base at Lake Biscarosse. After leaving Ford, the pair of Mosquitos crossed the sea at zero feet, making landfall at St Marcauf. They then flew Avranches-Laval-Angers-Les Sables, left the coast, climbed to 4,000 feet, flew south and parallel to the coast until opposite Biscarosse, turned port, dived to sea level and crossed the coast at the south end of the lake. Scherf turned left, Johnson went right, and the attack began.

Jim Johnson saw what he thought to be six anchored Arado 196 floatplanes about 800 yards (metres) east of the main slipway and he attacked, seeing cannon and machinegun strikes on two of the floatplanes. At once, light anti-aircraft fire, with red tracer, came up at them and Johnson broke violently starboard in a climbing turn, avoiding the fire.

Charlie Scherf saw a large pale blue flying boat (probably a BV222) on a slipway amongst the trees and tried to turn to attack it but it was too close, inside the radius of his turn; he fired but saw no results as intense light flak opened up on him as soon as he began the attack. He broke port but there was a severe jolt and a red flash below the port engine, later found to be from a cannon shell hit on the propellor.

Clear of the danger area, Johnson called Scherf on the radio, and got the reply, 'God, Johnny, I never thought I'd hear from you!'[3]

The aircraft then flew to Casanx and Bordeaux/Merignac, saw no activity, and returned to base at 20.00 hours. It was a two hours 25 minute trip, the return not made easier with the holed propellor and darkness. A visiting Group Captain sat in on the debriefing, and saw fit to rebuke Scherf for not turning back for a second attack at the flying-boat. Johnson recalled that, 'fortunately, my navigator was a big man, and got Scherf out of the room before a scuffle could occur.' Johnson found the operation interesting for three reasons: first, it was the first long range low-level daylight intruder mission planned and executed by 418 Squadron; second, he had never been more frightened in his life; third, the results of the attack — which came to light some 35 years later, when a scuba club in the area investigated the scene, verified the destruction of two aircraft, brought one to the surface, and identified it as a Dornier. The Squadron recorded this as 'an interesting experiment', with the description: 'Cloud cover was excellent and no opposition was encountered by this daylight hedge-hopping trip, until the target was reached.'[4]

By the end of November, Scherf had flown 20 operational sorties and had 492 hours. On 3 December he flew to Gotha-Erfurt-Leipzig, noting that the 'Target burned well. Weather

Charles Scherf DSO DFC

Flight Lieutenant Charles Scherf RAAF. (Beryl McLachlan)

u/s'. The flying conditions were such that there were no navigational pinpoints visible to or from the target—Brandis— and they had to penetrate a cold front and heavy rain. Like so many hundreds of others, the Lancaster flown by Scherf's friend Morrison was shot down on a raid to Frankfurt on 20 December. Morrison survived, was hidden by the Resistance and finally arrived back in England in May 1944. He was given several options, so decided to go back to Australia and join one of the B24 Liberator heavy bomber squadrons being formed by the RAAF for service in the Pacific.

Charlie Scherf's physical and mental attributes, plus his love of sport, allowed him to fit in well with the equally sport-loving Canadians. Jim Johnson, a good marksman, shot skeet with Scherf, recalling that Scherf's 'reactions were so fast that he could pulverise the bird before I pulled the trigger. I played tennis with Scherf, but it was no contest; he was so good.' Hal Lisson said that Scherf 'was a very good athlete, particularly in rugby. Ford had a very good side, but needed practice. 418 made up a team of Canadians plus Charlie, and acquitted ourselves quite well.'

However, sportsman, squadron and Mess member that he was, Scherf was still an Australian countryman, with firm opinions about dogs as pets and as working animals. The RAF Officers' Mess at Ford began to adopt stray dogs, and to allow the dogs into the Mess itself. Soon there were dog fights in the bar, and dog droppings in the ante-room. Enough was enough. One night, Scherf had enough to drink, and as Jim Johnson recalls, 'literally and physically kicked every dog out of the Mess. There were protests, but the message got through; the dogs remained outside. I quote him: "I've owned better dogs than they ever saw."'

On 2 January, Scherf and Brown flew to Ault-Abbeville-Poix-Amiens-Montdidier-Rosiers-Roye, again through bad weather, saw a twin-engined aircraft but lost it in cloud, which was 'broken stratus at 1,500 feet with rain in clouds. Increased to 10/10 south of Poix and in Montdidier area with heavy showers.' Intruder missions were affected by the weather, and the number of sorties reflected the difficulties of operating over Europe in adverse conditions. In January 1943, only 111 intruder sorties had been flown, with two enemy aircraft claimed as damaged. But in June 340 were flown, with 12 aircraft claimed destroyed. In November and December numbers declined to 271 and 100 respectively, with a total of five enemy aircraft destroyed.[5]

On 4 January, Scherf flew a dawn Ranger to Mont de Marsan and Biscarosse, then patrolled the Bay of Biscay but nothing was seen until returning to England, when he attacked a 200-ton ship with a seven-second burst of cannon and machinegun, seeing strikes, but no return fire. Similar flights were made in the next few weeks, such as one to Luneberg on the 21st, though visibilty closed in to 1,000 yards (metres) and with 10/10 cloud at 1,000 feet. On 8 January, squadron command passed to Wing Commander Don MacDonald,

Squadron Leader Hal Lisson. Flying Officer 'Fin' Finlayson and Lieutenant 'Lou' Luma.

with Paul Davoud being promoted Group Captain, posted to 22 Wing and awarded the DSO, the first such decoration for 418 Squadron. Soon after, Hal Lisson took over A Flight (from MacDonald) and Charles Scherf was given command of B Flight, when Massey Beveridge finished his tour. The squadron was soon to begin a period of high scoring against the Luftwaffe. On 21 January, the crew of Lieutenant Lou Luma USAAC and Flight Lieutenant Eckert destroyed a Messerschmitt 210 near Steinhuder Lake, bringing home two fragments of the Messerschmitt in their wing. Then came what 418 Squadron called 'a great day'. On 27 January, two formations of two Mosquitos flew a daylight Ranger operation over France. Wing Commander MacDonald, with Scherf, flew to Tours-Bourges-Avord, while Johnson and Caine went to Bourges-Clermont-Ferrand. It was a three hour 40 minutes flight, and generated great excitement in 418 Squadron: claims were for seven destroyed and one damaged.

Macdonald and Scherf had found nothing at Tours and went on to Bourges, where Scherf saw a He111 about ten miles (16km) ahead, and radioed MacDonald. The two Mosquitos, each armed with four 20mm cannon and four .303inch machineguns, leapt after the Heinkel. Macdonald came up on it from astern and below, opening fire at 175 yards (metres) range, shooting pieces off the Heinkel and setting the starboard engine and fuselage on fire. MacDonald broke starboard under the bomber, which almost hit him in its dive, and as Scherf went past he fired and hit the port engine, which also burst into flames. The unfortunate Heinkel crashed and exploded at 16.45 hours.

To avoid passing over Bourges airfield, with its flak, Scherf broke hard starboard, and Brown saw another aircraft ten miles (15km) away, coming head-on, south-east of Avord. With MacDonald now following, Scherf attacked. He curved to port, then again went hard starboard to come in behind the aircraft, identified as a four-engined Focke Wulf FW200, flying at 150 feet. Scherf attacked from behind, with slight deflection, firing a 10-second burst from 500 yards (metres). Closing to 200 and seeing strikes all over the port wing and fuselage, he increased his deflection, and hit the starboard wing and engine, then the starboard inner engine and fuselage began burning as Scherf broke off his attack. MacDonald had been filming the attack with his cameragun, and now fired a three-second burst into the port side and wing of the big Focke-Wulf, seeing one of the crew shot as he tried to bale out. Nose and wing down, the FW200 hit the ground and cartwheeled into a wood, burning furiously. It was 16.30 hours.

Scherf had gone to port, and again Brown saw an enemy aircraft, five miles (8km) ahead, east of Avord. Scherf called MacDonald and started after this enemy, but 'in the excitement, I had neglected to change over fuel cocks, and my port engine stopped.' MacDonald passed him and attacked the

Captured Heinkel He177 seen in Allied markings. Charles Scherf destroyed this type of aircraft. (Ken Merrick)

aircraft, a He177, flying at 1,000 feet. Once more coming in from astern and below, he fired a three-second burst, pieces flew off the starboard wing and fuselage and the starboard engines burst into flames. One man baled out and the big Heinkel spun down and exploded at 16.32 hours. MacDonald had filmed the entire action.

Meanwhile, having re-started his port engine, Scherf noticed a Ju88 approaching about 1,000 feet above, from the left, and thought it was about to attack, so climbed towards it, at the same time noticing the burning Heinkel going down vertically, but then the Ju88 slipped into cloud, and was gone. Scherf and MacDonald spoke on the radio and agreed to go home, at tree-top level, which was done with what Scherf called 'superb navigation on the part of my observer, F/O Brown'.

They were the first pair back, and the excitement of their successes was settling when Johnson and Caine returned claiming two Ju88s and two Ju34s destroyed plus a Ju86 damaged at Bourges and Clermont-Ferrand. The squadron was triumphant and the Officers' Mess celebrated with drinks for all plus a visit to the festivities in the Sergeants' Mess, where MacDonald and Brown made speeches of congratulation to the groundcrews. Assessing the operation, Scherf said that he was convinced that success depended on 'extremely low flying at high cruising speed to achieve the necessary degree of surprise,' as enemy aircraft were then silhouetted against the sky and became easier targets. In addition, the low flying height made it difficult for radar to detect the intruders or for ground observers to plot them efficiently. It had been his 28th operational sortie.

However, on 28 January, it was again brought home that war inflicts losses on both sides, when Flight Lieutenant Tommy Dubroy failed to return.

The grim process of intruding went on, and Tom Anderson flew his first, uneventful, operational sortie. On 3 February,

Anderson flew his second operation. With Flying Officer Frank Cadman as navigator, he set off to fly Avranche-Angers-Tours-Orlean. He saw the airfield Visual Lorenz lighting and an aircraft landing at Tours, so waited nearby. Sure enough, a few minutes later another aircraft came in along the Lorenz, Anderson followed it and fired cannon and machineguns at 200 yards (metres) range, 400 feet altitude. He saw flashes on the German, and broke away as the lights were doused and the airfield defences opened up. Anderson and Cadman could see the 'aircraft burning beautifully on the runway, and could still see it burning when 20 miles (32km) away.' On other operations, the crews of Luma/Finlayson destroyed a He177, and Kipp/Huletsky two Me410s before Scherf scored again.[6]

On 14 February, with Warrant Officer R.J. Gurnett replacing Brown as navigator, Charles Scherf flew as far as Lyon on the way to the designated target area of Marseille, but turned back; the cloud base was at 2,000 feet, and it was 'snowing like blazes'. Even so, it was a flight of three and a half hours duration.

Five nights later, taking off at 02.03, en route to Leipzig, Scherf was again accompanied by W/O Gurnett. At 03.12, near Florennes, Gurnett saw a single-engined aircraft with a bright amber navigation light on, heading south, and Scherf went to investigate. After a 20 mile (32km) chase, the aircraft climbed to 7,000 feet, and began turning tightly. Below was a white beacon, with 'full navigation lights', near St Hubert. The enemy aircraft began climbing while still turning, and Scherf closed in to attack, but when he was some 400 yards (metres) away, and about to fire, the single-engined aircraft dived. Scherf went after it, speed built up to 340mph (570 kmph) and both Scherf and Gurnett were 'astonished but delighted to see him pull out at very low level and crash, bursting into flames. We had not yet fired a shot.'

Scherf circled, filmed the wreckage and fired a two-second burst at it 'for good measure' then continued on to the target, which was 'burning nicely'. Cloud was down to tree-top level in the area. While returning they were followed by an unidentified enemy aircraft which dropped brilliant amber flares for some 60 minutes, but nothing else happened. In the 418 Squadron ORB, someone noted after the account of the destruction of the single-engined enemy, 'economy minded Charlie Scherf. Destroyed E/A (enemy aircraft) without firing a shot.'

That night, 79 RAF Bomber Command aircraft were lost

Squadron Leader Charles Scherf DSO DFC* RAAF. (Beryl McLachlan)

on the Leipzig raid. 418 Squadron's score of enemy aircraft destroyed stood at 35.5. The squadron crews were informed that their tour of operations had been increased from 30 to 35 sorties.

On 21 February, Scherf again accompanied Wing Commander MacDonald on a day Ranger, returning to the area where they had been successful on 27 January. They took off at 14.40 from Ford, crossed the Channel and at Cabourg went down to zero feet. At first, visibility was two to four miles (3-6km), with 6/10 cloud at 2,000 feet, but south of the Loire River cloud cover was total and there were snow storms.

There were no Luftwaffe aircraft airborne, and the speeding Mosquitos passed Heinkel He177s and single-engined aircraft on the ground at several airfields, but there were no worthwhile targets. As they approached Dijon, Scherf decided to strafe it, so pulled up to 300 feet and dived on a He111 on the south-east edge of the field. He fired a two-second burst with all guns, saw strikes on the cockpit area, and passing the starboard wing of the Heinkel at zero feet, fired at dispersals on the northern side of the airfield, flew past more Heinkels on his left and set off home. On the way, he saw a wireless mast, and a large swastika flag flying, the red-white-black colours standing out in the gloom, with what seemed to be a large group of Germans at some sort of parade or ceremony. He dived and let fly into their midst with cannon and machineguns, then was past and away. However, they had hoped to shoot down more aircraft, and Scherf made the notation in his log book, 'Wing-co very disappointed.' Returning to England from another sortie, he came on a damaged US bomber, which he escorted over the Channel and to an airfield. Scherf landed first, climbed out and watched as the B17 made its approach. It undershot, hit the ground, exploded before their eyes. Its throttle quadrant flew through the air and landed at his feet. He kept it and gave it to his sister, Beryl.

On the 24th, he was again flying at night, a 'Flower' operation, patrolling between Ansbach and Illesheim to disrupt Luftwaffe nightfighter flights, with Flying Officer Finlayson in the right-hand seat. The Bomber Command target was Schweinfurt. Scherf took off at 23.30 and began covering the area between four enemy airfields. After about 45 minutes, at 02.28, they saw a large enemy aircraft with a white tail light going in to land at Ansbach. Scherf wheeled after it, closing fast, when the German switched on his landing lights, which reflected in the haze, making it difficult to estimate the distance for firing. The range was closing quickly and Scherf passed a mere 20 feet (6m) below him, identifying it as a four-engined Junkers Ju290.

As they circled to go in again, they saw the big Junkers had landed, and was rolling along the runway with navigation lights on, as were two other aircraft. Then they saw a fourth aircraft, with full navigation lights, coming in to land, so Scherf swung in behind it, height 700 feet, range 200 yards (metres), fired two seconds of cannon, and immediately saw strikes and fire from one engine. They then passed below the target, recognised it as a Ju88, and watched it crash and burst into flames. It was 02.30. Scherf swung out to the edge of the airfield, into the darkness, and there was yet another aircraft, lights on, coming in to land. Scherf banked around to come in astern, applied some deflection and fired a five-second burst, saw no result, increased deflection and fired for three seconds. The strikes flashed, a fire started and the aircraft went down to burn on the edge of the airfield at 02.39. All Scherf and Finlayson could see was that it was a twin-engined aircraft; type identifcation was not possible. They continued patrolling for 15 minutes, then returned to base. It was Scherf's 33rd operation.

By contrast, Tom Anderson was on his fifth operation, to Coulommiers-Melun-Bretigny. His generator failed, leaving him without radio or lights, and he had to return using a flashlight for cockpit illumination; then the throttles stuck open as he was landing. He noted it as a 'shaky do.' Flight Lieutenant Don McFadyen destroyed a Messerschmitt Me410, and another enemy aircraft was destroyed by the Williams/Gurnett crew. Next day, 25 February, Wing Commander Dick Bennell arrived to take command of 418 Squadron, and Hal Lisson, tour expired, was posted. Air Marshal Roderic Hill, Air Officer Commanding Air Defence Great Britain (AOC ADGB) visited the squadron. Tom Anderson flew his sixth sortie, dead reckoning in cloud all the way to Biblis, north of Mannheim. They saw three Visual Lorenz and 'about 100 searchlights', finally returning to base unsatisfied. However, Flight Lieutenant Johnnie Caine destroyed a Messerschmitt Bf110 on the ground, raising the squadron score to 40.5.[7]

Scherf's next operation, on 26 February, was another day Ranger to the area of Bourges-Tours-Dijon, with Finlayson, and accompanied by Flight Lieutenant Cleveland, who had Flight Sergeant Day RAF as his navigator. It was to be the first in a series of epic sorties.

It was a four and a half hour trip, and they left at 15.20, heading across the Channel to France. Making landfall at Point de la Percee, they sped along at tree-top height without sighting airborne enemy, until they reached the airfield at St Yan. The countryside was flat, and they saw parked aircraft some distance away. They moved into line abreast and Scherf went for two Ju52s on the left, while Cleveland attacked a He177 sitting out in the middle of the field. Groundcrew were working on the bomber, and the low-flying Mosquito was on them before they knew anything was wrong. Cleveland's first burst hit the ground in front of the big Heinkel, sending the mechanics scattering. He lifted his nose slightly and the hail of fire moved up onto the bomber, strikes sparking, hits on the wingroot, and a large flash shot out of the plane. Then he was passing over the burning Heinkel, no more 10 feet (3m) above the tail, coming around to starboard.

Meanwhile, Scherf had climbed to 500 feet and was swiftly scanning the target area as he approached the Ju52s, roaring over the heads of a group of about 50 Luftwaffe groundcrews, easily visible in their dark overalls, who were all walking across the field, some of them looking up at the fast twin-engined plane overhead. Finlayson looked over to starboard and saw smoke pouring from Cleveland's engines, and opened his mouth to tell Scherf that Cleveland had been hit and was going down, then realised that it was caused by the extra boost when Cleveland opened the throttles to attack.

Scherf opened fire at 600 yards (metres), with all guns, seeing flashes on one Ju52. Then, as he closed, the other burst into flames and he was on them, pulling up through the smoke and fire. Finlayson looked back and saw Cleveland's Heinkel burning, and as Cleveland was passing he watched the two Junkers with spreading fires on them, assisted by a petrol tanker also in flames. Some light ground fire was ignored by the Mosquitos as they hurtled across the airfield, and it was 17.20 hours.

Scherf looked at the counter to see how much film had been used, and saw that the camera had not been turned on. They set course south for 20 miles (33k) to fool any German observers, then turned back north-east to Dole. However, they missed the airfield there because they passed at low-level on the wrong side of a hill, and spent about 15 minutes trying to find it again. Navigation was not made easier as the rivers were in flood, and were not always as shown on the map. However, Finlayson guided them back to Dole, and just as they saw the airfield, there was a Heinkel 111Z towing two Gotha Go242s gliders in line astern. The Mosquitos attacked.

Scherf afterwards called the Heinkel 'The Monstrosity',

German Gotha Go242 glider — a type destroyed by Charles Scherf. (Ken Merrick & David Vincent)

Heinkel He111Z — two He111 joined. Charles Scherf and a Canadian pilot destroyed one. The squadron unsuccessfully requested each pilot be awarded a victory since there were two fuselages. (RAAF official)

and that is how it compared with most other aircraft of the day. It was the result of an effort by the German aircraft industry to provide towing aircraft for the very large gliders being built at the time, and consisted of two He111 fuselages and outer wings joined by a central span with three engines, making a total of five for the machine.

The Mosquitos surged towards the Heinkel, which was at 2,000 feet, but flying slowly. Scherf found that he was overshooting, so broke off his climb, while watching Cleveland fire from 300 yards (metres) in to 50 yards (metres), and Cleveland almost hit the rearmost Gotha as the tow-rope broke, the Gotha's nose went straight up as the strain was suddenly released, then pitched straight down, the pilot's problems probably exacerbated by the slipstreams of the Heinkel and passing Mosquito. He failed to regain control and Scherf saw him crash into a house and hedge directly below.

Scherf pulled his Mosquito around onto the lumbering target, and had the cameragun running as he came in to fire as the second Gotha released from the tow, but his two-second burst disintegrated it and he was in danger of colliding with pieces of the glider as it fell. He passed the fragments and was on the Heinkel, firing a three-second burst which brought fire from the starboard engine as he passed, with Cleveland again in position behind. Cleveland fired cannon and machineguns from 300 yards (metres), closed to 100, watched his shot striking the engines and fuselage on the starboard side, then broke away. Scherf was there again, hitting it with machineguns, as his cannon had been used up. Cleveland was a little surprised at the Heinkel's ability to absorb punishment, as he had fired about 10 seconds of cannon into it, yet despite 'the fact that three engines were burning, and bits of wreckage flying everywhere, the Heinkel seemed to take a very long time before it started a slow spiral to the ground,' with the

Mosquitos following. Port wing down, possibly in an attempt to keep the flames away from that fuselage, as the starboard one was flaming, the ungainly machine hit the ground and burned. The victors flashed overhead, and Cleveland fired more machinegun at another Gotha glider parked in a field, saw strikes on the nose, and then they were past and gone.

The mission resulted in claims for the He177 and two Ju52s at St Yan destroyed, plus the Heinkel 111Z and two Gotha Go242s destroyed and another Go242 damaged. In a brave attempt to increase the score, the squadron asked that Scherf and Cleveland each be credited with a He111 destroyed, as the He111Z was really two joined.

This was Scherf's 34th operation, and he had a total of 606 hours; his tour with 418 was to end, and he wrote in his log book, 'Last trip with a great squadron', but later added, 'I thought'. At the end of February, his log book was endorsed with an 'Exceptional' rating as an Intruder pilot, and the comment, 'An exceptional pilot with outstanding fighting qualities.'

Posted to a staff job at what was earlier Fighter Command, but in early 1944 clumsily titled 'Air Defence Great Britain' (ADGB), Scherf, like many active combatants before and since, found the daily life and routine boring and galling. As intruder controller, his job was to monitor the plotting table for enemy activity, and to maintain the status board for the availability of intruder squadrons. If the Luftwaffe was attacking England, intruders were sent to cover the German bomber bases, in an attempt to catch them on return, but when RAF Bomber Command was attacking targets in Occupied Europe, the intruders were sent to patrol Luftwaffe nightfighter bases.

In the 30-day period to 26 February, 418 Squadron had destroyed 24 enemy aircraft, while in the previous 22 months of its existence it had destroyed a total of 22½. A combination of experience, suitable aircraft, tactics, enemy activity and opportunity to engage was producing these results. Wing Commander MacDonald was posted from 418 Squadron and Wing Commander Dick Bennell DFC assumed command of the squadron. Also tour-expired was Squadron Leader Hal Lisson, succeeded as A Flight commander by Howie Cleveland, with Bob Kipp taking over Scherf's B Flight. Tom Anderson flew Scherf to Northolt in the squadron Airspeed Oxford on 7 March. The squadron still was scoring victories. The night before, Flying Officer Herb Jones was hit by friendly fire from an RAF bomber, which scored eight hits, one in the cockpit, starting a fire there; he returned safely. Lieutenant Lou Luma USAAC was looking for aircraft around Pau, near the Spanish border. He saw one making a night landing, closed on it, and identified it as a 'long-nose' Focke-Wulf 190D — a very dangerous opponent. Luma exploded it with a three-second burst, but debris damaged the starboard radiator on the Mosquito, and Luma flew 600 miles (960km) back on one engine, at 190mph (300kmph).

The dangers of operating as intruders were brought home even more strongly two days later, on 9 March. On the first trip of his second tour, a day Ranger with Johnny Caine and Earl Boal, Wing Commander Dick Bennell, with navigator Flying Officer Shields, failed to return. Caine had no warning of trouble, and it was presumed that Bennell had hit the ground. Hal Lisson, with nine months service on the squadron, and having just finished a tour of operations, assumed command.

Over the Continent, anti-aircraft defences became ever stronger, and on one occasion Flight Lieutenant Don Mac-Fayden was trapped and held in searchlights, despite all his evasive manoeuvres, which included going down to below tree-top height; the searchlights could not be shaken off.

More aircraft were returning with flak damage from the operations.

On 18 March, the squadron was formally titled 418 (City of Edmonton) Squadron, and became a ward of the municipality of Edmonton, Alberta. Squadron operations continued,

Flight Lieutenant Don MacFadyen and Flying Officer 'Red' Stewart. (418 Sqn RCAF)

and on 21 March Don MacFayden and Lieutenant Luma USAAC flew a day Ranger to Hagenau, destroying three German aircraft in the air, four on the ground, and damaging twelve more. Tom Anderson had another 'shaky do', when all his gyrocopic instruments became unserviceable just after he crossed the Elbe River going east. He managed to return to Coltishall, at night, on compass. Wing Commander Tony Barker, with his navigator Flight Lieutenant Gordon Frederick, arrived on 30 March to take command of the squadron.

In March, Scherf did some flying in Spitfires, a Typhoon and other more sedate types, but by early April found he could not keep away, and went back to 418 Squadron at Ford,

Bill Stewart RCAF, Charles Scherf's navigator on some of his successful flights. (Mary Stewart)

for another day Ranger over France. In his log book, he wrote, 'Very brassed off at ADGB, so went on a day Ranger — Ford.' He had arranged a 48-hour leave from Intruder Control at ADGB. As Scherf was a friend, and had already arranged for a navigator and accompanying crew, Hal Lisson reluctantly agreed to the operation. With Flying Officer 'Red' Stewart as his navigator, and Flying Officer Caine in the other Mosquito having Pilot Officer Boal with him, Scherf took off at 15.57 hours for the old hunting grounds in France. Stewart had already done 25 sorties but had not seen any German aircraft destroyed in the air. Stewart's usual pilot had been injured in a motorcycle accident so he was available to accompany Scherf.

When they reached the French coast, the cloud cover was poor, leaving them visible to attack from above, so Caine turned back. Scherf went on to Lyon, crossing into France at Point de la Percee and staying at tree-top height. Near Monsereau a Fieseler Storch observation aircraft was seen, flying west. Scherf throttled back to 150mph (250kmph) and attacked from the rear, cameragun operating, but to Scherf's surprise his cannon would not fire, and he passed the helpless Storch, breaking around to come in again. He found that the cannon switch was not fully down, adjusted it, and prepared to destroy the frail observation plane. But the German pilot had seen him and done the sensible thing: landed and, with his passenger, was running for the woods. The Mosquito swept down, and blew up the Storch with a short burst at 17.02. Scherf passed over and away, looking for more prey. There was none to be found. He flew to Limoges, Clermont and Rhue, but there were no Luftwaffe aircraft to be seen. Then, at 18.05, he was approaching Lyon from the south and saw four aircraft flying inside the airfield area.

With the camera going, he flew at one about 1,000 feet above the ground, noticing the twin braced tail, and gave it a two-second burst from 200 yards (metres). It blew up and fell burning to the ground. It was 18.15. Scherf continued to circle the airfield, and another aircraft, with twin engine and single fin, came to investigate the smoke and flame. Scherf fired another two-second burst at 200 yards (metres) and this pilot 'paid for his curiousity by being sent down in a ball of flame (18.19).' During these combats, the other two aircraft, possibly flown by more discreet pilots, had landed, but another was still circling over Lyon itself. 'In view of the flak defences of the city,' said Scherf, 'we decided to let well alone and flew north-east to Amberieu.'

After investigating the empty airfield there, he flew into the mountains to wait until dark for the return across the French coast to England, whiling away the time by photographing a mountain resort, 'for post-war advertising purposes.' But they decided to set off for home in daylight, and flew to St Yan.

There they saw two He111s parked on the south side of the airfield, and attacked at a height of 50 feet (15m). The four-second burst of cannon and machinegun fire brought instant results, with flame and explosions as the Mosquito went past. Scherf claimed the Storch and Heinkels destroyed on the ground and two unidentified twin-engined aircraft destroyed in the air. Later these were identified as a Messerschmitt Bf110 and a Focke-Wulf Fw58. He noted in his log book, 'Pleasant day.'

Temporarily assuaged by the latest operation over Europe, Scherf went back to ADGB, keeping his hand in with Spitfire flights during the rest of April. However, the reports of 418 Squadron exploits arriving at the headquarters did nothing to quieten him; indeed, they probably resulted in the opposite. On 14 April, Squadron Leader Bob Kipp and Flying Officer Johnnie Caine went to Denmark, destroying seven and damaging two more Junkers 52s and Dornier 217s, then outran two

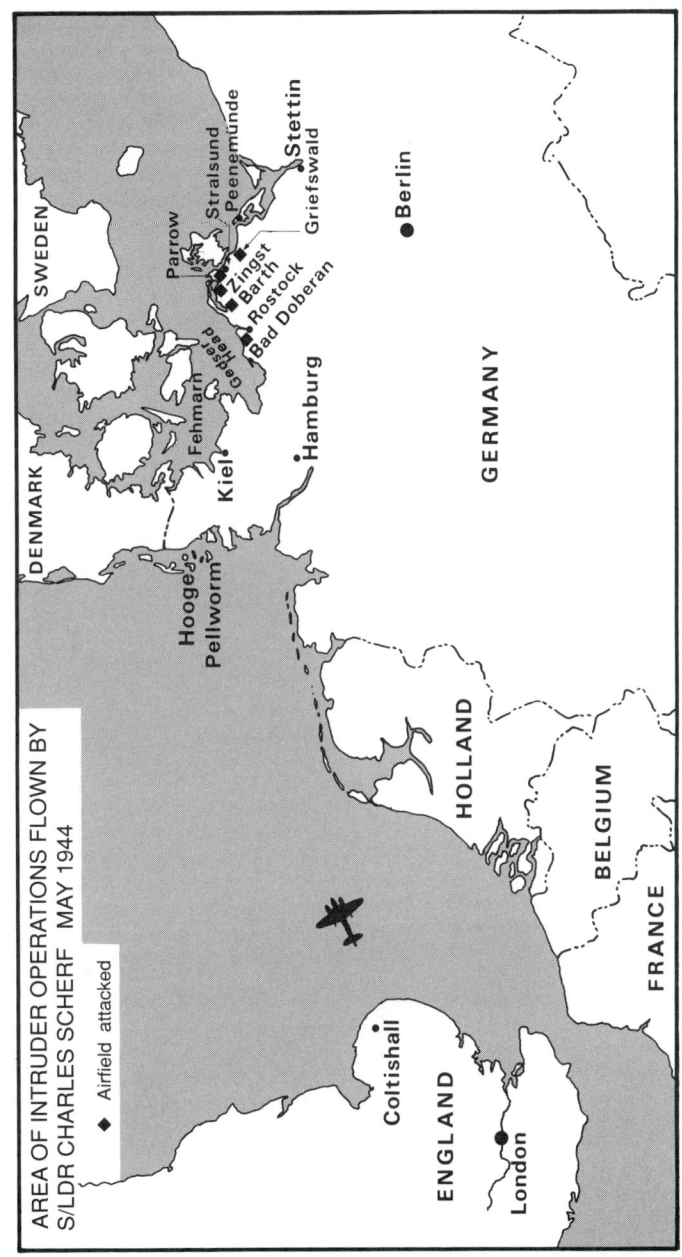

long-nosed Focke-Wulfs. Two days later, Squadron Leader Howie Cleveland, Flight Lieutenants Kerr and Harper and Flying Officer Jasper, with their navigators, flew Rangers in the Luxeuil and Metz areas, destroying 10 more enemy aircraft. Flying Officers G.N. Miller and G.D. Miller destroyed two, claimed two more as probables and damaged four other Luftwaffe aircraft.

On 2 May, Scherf flew an Oxford from Northolt to 418's new base at Holmsley, and was given a Mosquito. Again with 'Red' Stewart in the navigator's seat, and Caine and Boal flying as No. 2, at 14.30 they set off from Coltishall for a different area — North Germany. There was rain and low cloud throughout, good weather for intruders seeking aircraft with their pilots concentrating on other things.

The Mosquitos made landfall at Hooge Island, crossed the peninsula and flew south along the south to the seaplane base at Ribnitz. Scherf saw at least 15 Heinkel, Dornier, Blohm und Voss and Arado seaplanes moored below and ahead, picked two He115s and attacked. He closed on them and fired

Blohm & Voss BV222, a type strafed by Charles Scherf in France, 1943. (Ken Merrick)

a five-second burst, watching many strikes flashing and sparking, but there was no flame or explosion, as probably the fuel tanks were empty. At the same time, Caine had fired on a group of four Dorniers, opening at 500 yards (metres) and closing to 50 yards (metres). All four targets were hit, and two burst into flames then exploded, confirmed by Scherf and Stewart as they swung off the target area, setting course for Barth. It was 15.48.

At about 16.20 they reached Barth, seeing at least 12 enemy aircraft on the ground. Scherf strafed and blew up a He111 and a Dornier Do217, while Caine destroyed a Ju34, a Ju52, a Ju86 and a Ju88, the Mosquitos making a series of passes from many directions back and forth across the airfield. Caine flew through the flames of the exploding Ju34 and the explosion of the Ju86, some pieces of which hit his plane.

Leaving a scene of destruction behind, they headed for Griefswald, but five minutes later Caine was reduced to one engine when his port Rolls Royce Merlin stopped. Scherf was not in sight, and had drawn on ahead anyway. They discussed his predicament by radio, and Caine decided to try for Sweden. However, soon after turning on a northerly course, he changed his mind and turned again for England, eventually arriving back at 20.06, after Scherf.

In the meantime, Charles had gone to Griefswald, where an unlucky Ju86P was flying at 1,000 feet. Scherf sailed in astern and fired a one-and-a-half-second burst at 300 yards (metres), sending the Junkers down in flames, while Scherf turned steeply right over the airfield to bring his guns onto a He111, one of 10 parked there. It blew up after a one-second burst, and the Mosquito was gone into the rain and weather, back to Ribnitz. On the way they passed over Bad Doberan, where Charles used the last of his ammunition in a one-second burst on yet another He111 which burst into flames. Flying over Kiel Bay, they saw a convoy of about 15 ships, some as large as 6,000 tons, entering the Kiel Canal, but no one fired at the lone twin-engined aircraft heading east. Later, limping back on one engine, Caine and Boal also passed over the convoy without incident.

Scherf claimed the Ju86P destroyed in the air, and the Do217 and three He111s destroyed on the ground, plus two He115s damaged, while Caine claimed two Do18s, a Ju34, a Ju52, a Ju86 and a Ju88 destroyed on the ground, plus two Do18s damaged. These brought 418's score to 98 destroyed and 53 damaged. Charles noted it as 'a very pleasant day off from ADGB duties.'

Junkers Ju86, a type destroyed by Charles Scherf. (Ken Merrick)

When asked where Caine was, Scherf had said he was by now in Sweden, but a Mosquito on one engine appeared in the circuit at Coltishall and landed, Caine having changed his mind. Elsewhere on the night of 2 May, Squadron Leader Bob Kipp, on a sortie in the Munich-Saarburg area, destroyed four FW 190s, but debris hit his Mosquito, it stalled, and he recovered at only 100 feet, while Don MacFayden destroyed another He111Z and damaged a twin-engined aircraft, returning on one engine from Dijon. The squadron score had increased by sixteen in 24 hours, to total 103.5.

By this time, the newspapers were reporting each of these 'extra' sorties, and all those who took part were receiving a certain amount of publicity in national and local papers. Of course, Scherf's reputation was spreading in the intruder squadrons and RAAF as well. For the younger less successful pilots, who saw him in the Mess after these sorties, he was regarded as 'fun-loving, gregarious, and central to any group around.' John Howell 'admired him tremendously. Having read so much of Charles Kingsford-Smith, I can see the faint likeness of a dare-devil spirit, matched with a competence and killer spirit.'[8]

On 11 May, Wing Commander Tony Barker DFC and his navigator, Flight Lieutenant Gordon Frederick, ditched in the Channel but were picked up by air-sea rescue launch after 18 hours in the dinghy. Next day, the Squadron celebrated the safe return of Barker and Frederick, the award of the DFC to Bob Kipp and the squadron score reaching the century — 100 enemy aircraft destroyed. This was all part of the life for which Charlie Scherf yearned.

Meanwhile, in the wider war, apart from Japanese offensives in Burma, the Axis forces were on the defensive everywhere. In Russia, Italy and the Pacific the Allies had made more advances, destroying, surrounding or by-passing large numbers of German and Japanese troops. For the Germans, the demand for unconditional surrender had left them with the simple choice of fighting on to the bitter end, whatever the cost and sacrifice to their nation, or surrendering to accept whatever treatment the victors wished to inflict. With President Franklin

Roosevelt allegedly surrounded by Jewish advisers, and vivid memories of the punitive demands on Germany made by the Versailles Treaty after the defeat in World War 1, the average German believed that there was little choice. In any case, the firm grip exerted by the Nazi Party and various police forces allowed little opposition to Hitler. To the Japanese, ruled by a divine Emperor, surrender was literally unthinkable.

On 16 May, Charlie Scherf was back at 418 for yet another day Ranger along the German Baltic coast. With Finlayson beside him on this 37th operational sortie, and Squadron Leader Cleveland in the other Mosquito, they took off at 15.00, across the North Sea. As they approached Gedser Head, an aircraft was seen ahead, at 2,000 feet. They chased it, the aircraft opened up his engines, leaving a black trail, and dived to 800 feet. After five minutes, Scherf was 150

The formidable Focke Wulf FW190. One was shot down by Charles Scherf in a Mosquito. (Ken Merrick)

yards (metres) from the fleeing He111, and opened fire with a two second burst that brought flames from the fuselage, starboard wing and engine, sending the Heinkel into a dive down to the sea, and it exploded on impact at 17.05.

The hunters went on to Zingst, arriving 20 minutes later. A Focke-Wulf FW190 was diving and climbing, practising rocket firing into a field east of the airfield. It was at 1,000 feet and Scherf climbed from the trees to attack. The 190 broke sharply port, but Scherf reefed the Mosquito around after him, firing a 30-degree deflection shot from 300 yards (metres) which hit and flames shot out of the Focke-Wulf; it straightened up momentarily: Scherf got in behind and fired a two-second burst from dead astern; the 190 flicked onto its back and exploded, the remains falling into the field below.

Scherf then led Cleveland out over Kubitzen Bay, intending to go on to Vitte and Bug, but noticed activity in the air over Parrow airfield, so swung that way. Cleveland veered left to Rambin, while Scherf went south to Parrow. At 17.30 he made a head-on attack from below on a He177, closed to 150 yards (metres) and delivered a one-second burst which exploded on the cockpit and engines, setting them on fire, and the Heinkel nosed down into the sea, with one man seen to bale out.

The Mosquito roared on, Scherf swinging to approach Parrow airfield from the east. He and Finlayson saw Cleveland attacking a twin-engined aircraft coming in to land, but could not watch the results as they were coming in on their own targets. At 300 yards (metres) a two-second burst was fired at a He111 parked by the slipway, and it exploded at once, 'the nicest explosion we have ever seen,' said Finlayson, noting the time as 17.32.

They broke north over the bay, calling to Cleveland that a biplane was going in his direction, but he replied that he could not attack it at the time, so Scherf went after it. At

17.34, at 500 feet, he closed to 50 yards (metres) and a one-second burst completely destroyed the Henschel Hs123. Off to the right they had seen about 20 flying-boats anchored at the Parrow seaplane base, so dived on them. At 17.35, a two-second burst from 200 yards (metres) produced strikes and smoke from a Dornier Do18, and immediately they turned after a Ju86 seen flying over the town of Stralsund. Scherf tried to avoid flying over the town, but accurate automatic flak, 37 or 40mm, came streaming up around them, hitting the Mosquito in the tail and on the port drop tank, making a vivid flash as it struck. Scherf went on through, attacked the Junkers and fired for two seconds at 200 yards (metres), both engines on the Junkers exploded and the burning plane hit the ground just south of the town. It was 17.37.

They broke starboard and headed north, away from the target area, noticing the Dornier still smoking. As they passed Zingst, the FW190 was still burning, and a Messerschmitt Bf109 was seen on the airfield, but they left it and went on. Ahead of them, on one engine, was Cleveland. They caught up with him near Darsser Point, and over the radio he claimed two enemy destroyed. Charles gave him a course to steer for Sweden, then flew on. Swedish anti-aircraft guns opened fire, the Mosquito was destroyed, Howard Cleveland survived, wounded, but his navigator, Frank Day, was killed.

Again over Kiel Bay, Scherf passed a convoy, and again there was no fire at them. But as they were leaving the German coast near Pellworn Island, a flock of small birds on the beach were alarmed and flew up into the path of the Mosquito; 27 hit the leading edge of the wings, nose and propellor spinners. The Mosquito flew on. Arriving at Coltishall at 19.36, it was found that most of the port elevator had been shot away, and 20 of the birds had struck the leading edge of the wing, being driven through it and on out through

Starboard outer wing of Mosquito, 418 Squadron, showing some of 27 bird strikes on 16 May, 1944, after which Charles Scherf flew it back to England from Denmark. (Hope Scherf)

the top. There is no doubting the sincerity of Finlayson's remark at the conclusion of the combat report: 'What a kite this Mosquito!'.

For this epic sortie, Charles Scherf claimed a He111, a FW190, a He177, a Hs123 and a Ju86 destroyed in the air, plus the He111 destroyed on the ground and the Do18 damaged on water. After assessment by Fighter Command, Cleveland was awarded a Do217 and He111 destroyed and an unidentified aircraft damaged on the ground. These brought 418's claims to 121 destroyed, 8 probables and 63 damaged in the air and on the ground.

It was Charles Scherf's last operational flight. Until he left the UK, he had to content himself with flights in a variety of combat and communications aircraft, until on 8 July he 'beat up ADGB' in a Spitfire, noting 'last flight in a great country.'

The intruder sorties with 418 Squadron were part of the great aerial effort in the lead-up to the D-Day invason. From the winter low of 100 sorties in December, February saw 230, rising to 389 in April and climbing again to 508 sorties for May. A total of 42 enemy aircraft were claimed destroyed, with others claimed as probables and damaged. The Luftwaffe was no longer safe anywhere in France, nor in eastern Germany or along the Baltic. Apart from the fighter escorts of the large US daylight bombing formations, RAF intruders were skimming along at tree-top level, attacking everything in sight. Luftwaffe records for France and Germany for this period are such that it is impossible to correlate them and RAF claims, but they do include reference to 54 destroyed in the air, 11 on the ground and another 11 damaged, all credited to intruders. Three weeks after Scherf's final foray along the Baltic coast, the Allies landed in Normandy, and once ashore the writing was on the wall for Hitler's Third Reich. After the fighting in the hedgerows, the Allied armies broke out and annihilated the German armies in France, beginning the advance to Berlin. It was only an agreement between Roosevelt and Stalin which allowed the Soviet hordes to take the German capital, where they remained for over 40 years with huge numbers of tanks and men.

Charles Scherf had been awarded the DSO, DFC and Bar. The first DFC was gazetted on 4 April 1944, the citation referring to his exceptional keenness, courage and determination, and destruction of four enemy aircraft at night. The Bar to the DFC was gazetted on 12 May, with reference to the sorties of February and April. The DSO was gazetted on 27 June, the citation referring to two daylight sorties in May, into Germany, adding, 'His successes are a splendid tribute to his great skill, enterprise and fearlessness. This officer has set an example of the highest order.'

Finally finishing operational flying, Scherf was guest of honour at a dinner in the Savoy Hotel, attended by many of the senior fighter leaders and headquarters officers, during which many compliments were paid him. Air Chief Marshal Sir Charles Portal, Chief of Air Staff, said in his speech,

> Charles, I want you to know I love you as a son, I admire you as a son, and I believe you are the most outstanding airman I have ever known. When you are in a plane you seem to be the soul of it.

In July 1944, when Scherf was at the embarkation depot at Wigan, he again met Morrison, his friend from training at Temora, back after being shot down and on his way to Australia. Again, they sailed the Atlantic, this time in the opposite direction, on the *Queen Elizabeth*. They arrived in New York and then, as Morrison recalled, 'Charlie calmly told me he was leaving the draft and going across to California to see his sister, and then hitch a ride to Australia on a US military aircraft. This was typical of Charlie Scherf, an independent man of action. This independence sometimes brought him into conflict with his superiors.' So Morrison and Scherf parted.

In New York and across the USA, Charles Scherf enjoyed the hospitality for which Americans are famous. In San Francisco he visited his sister Beryl, and on one occasion was asked if there was anything special he wanted for dinner. He asked for bully beef and white onion sauce, pineapple upside down cake and lots of ice cream. He was still unwinding from the stress and pressures of operational flying and, on one occasion, when a little drunk, climbed onto a table in a restaurant, with another airman, and they jived there. Such was the spirit of the times that no one objected, but applauded, and it was accepted as fun, by servicemen relaxing after combat.

Scherf returned to Australia, where his wife, Hope, was waiting with their two first children. He was posted to 5 OTU at Williamtown, near Newcastle. This was not to his

Squadron Leader Charles Scherf DSO DFC* RAAF. (RAAF official)

liking and he noted in his log book, 'Worse Luck.' On 19 October, in Mosquito A52-1006, as part of a 'showing the flag' cross-country, he went to Glen Innes, and noted: 'Bad Day. Now under open arrest for beating it up.' His sister Beryl believes that there was a certain amount of jealousy at the publicity given Charles during and after his tour of operations, particularly on arrival back in the Emmaville-Glen Innes district, and that there was vindictiveness in the complaints about his low-flying display disrupting the peace and upsetting hospital patients and the elderly. It was wartime, and Charles was duly punished by the RAAF, by being grounded—for one hour on a Sunday. He flew until March 1945, the last flight recorded being on the 15th, Richmond-Kingaroy, in Mosquito A52-516. Whether he continued to record all his flights is questionable and he had not been totalling his flying hours since completing his tour with 418 Squadron in February 1944.

The war ended, and millions of servicemen had to consider life in peacetime. Many had grown accustomed to the excitement of war, and many had seen friends killed or wounded. Adjusting to the routines of peacetime was easier for some than for others.

Mr Scherf senior was ill, so Charles went back to managing 'Big Ben', but it was obvious that he missed the air force. He had been offered a job by Juan Trippe, head of Pan Am airlines, but refused, returning to 'Big Ben' and his family.

418 Squadron ended the war, after 37 months of operations, having flown 3492 operational sorties, totalling 11,248 hours, destroyed 178 enemy aircraft in the air and 73 on the ground, plus nine probables and 103 damaged; 83 V1 flying bombs were destroyed. The squadron did not use airborne radar, all victories being achieved by visual contact. One hundred and forty three aircrew were killed, captured or listed as 'missing'. 62 decorations had been awarded squadron members, including

three DSOs, 43 DFCs and nine Bars to the DFC, five DFMs, one US DFC and one US Air Medal.

In 1947, Scherf's sister Beryl and her American husband, with their two children, flew to Australia by Pan Am DC-6 Clipper, quite the most modern way to travel. There was a large family reunion at 'Big Ben', and though Charles seemed well, she noticed a great change in him. He was quiet but restless. At night, he would talk to Beryl and her husband about the war in Europe, the toll of people he had known, and remarked that the Germans he had killed seemed to march across his bed in the darkness. His drinking increased, and he would drive his Oldsmobile at great speed, scaring the passengers. In a tragic accident, he shot his favourite dog, Sally, by mistake when hunting lamb-killing foxes. This mishap seemed to weigh on him very heavily.

Charles Scherf was killed in a car accident on 13 July 1949. 'When he died,' said Beryl, 'a light went out of our lives and we were never the same again.' His parents were heart-broken, and sold 'Big Ben', unable to face the prospect of Charles not continuing the work on it. Hope Scherf was left with four children and no assistance from the government except a widow's pension. It took intervention by Service friends of her husband, and legal action as far as the High Court, before a Service pension was awarded her. Squadron Leader Charles Scherf DSO DFC* was as much a victim of the war as those Luftwaffe crews he had shot down over France and Northern Germany.

At Charles Scherf's funeral, two Mustang fighters from RAAF Williamtown circled as the cortege left the church, then made a low pass as the mourners from the 40-car procession gathered at the graveside. As the service by Reverend Siddell was almost finished, more engine noise was heard, and a Mosquito from Archerfield appeared, circling

Charles Scherf DSO DFC

the cemetery, turning in to make a low pass with one engine stopped just as the prayer was ending and the casket was being lowered into the grave.

Four American members of 418 Squadron wrote to the Mayor of Glen Innes when they heard of Charles' untimely death. Part of their letter cannot be improved upon as a tribute to Charles Scherf:

> For sheer courage and determination we never met his equal, and in our hard and trying times he had that saving sense of humour.

Appendix

Sorties by Charles Scherf resulting in enemy aircraft destroyed or damaged.

Date	Aircraft	Result	Location
28 November 1943	Mosquito R	one BV222 damaged	Biscarosse
27 January 1944	Mosquito R	one FW200 destroyed	Bourges
19 February 1944	Mosquito R	one ? destroyed	St Hubert
21 February 1944	Mosquito J	one He111 damaged	Dijon
24 February 1944	Mosquito R	one Ju88 destroyed one ? destroyed	Ansbach
26 February 1944	Mosquito F Mosquito F Mosquito F	two Ju52 destroyed one He111Z destroyed one Go242 destroyed	St Yan Dole Dole
5 April 1944	Mosquito F Mosquito F Mosquito F Mosquito F	one Fi156 destroyed one Bf110 destroyed one FW58 destroyed two He111 destroyed	Monsoreau Lyon Lyon St Yan

SIX ACES

2 May 1944	Mosquito X	two He115 damaged	Ribnitz
	Mosquito X	one He111 destroyed	Barth
	Mosquito X	one Do217 destroyed	Barth
	Mosquito X	one Ju86P destroyed	Griefswald
	Mosquito X	one He111 destroyed	Griefswald
	Mosquito X	one He111 destroyed	Rostock
16 May 1944	Mosquito T	one He111 destroyed	at sea
	Mosquito T	one FW190 destroyed	Zingst
	Mosquito T	one He177 destroyed	Parrow
	Mosquito T	one He111 destroyed	Parrow
	Mosquito T	one Hs123 destroyed	Parrow
	Mosquito T	one Do18 damaged	Parrow
	Mosquito T	one Ju86P destroyed	Stralsund

Pulled up to clear the mast

Tasmanian Whirlwind

Plt Offr Max Cotton DFC

SIX ACES

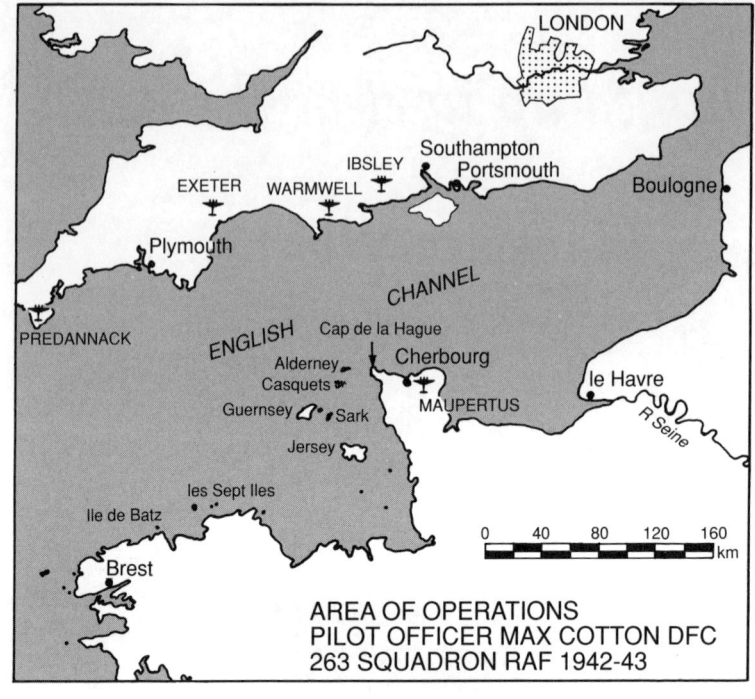

Max Cotton DFC

Pilot Officer Max Cotton DFC RAAF was a young Tasmanian pilot who served on 263 Squadron RAF, flying Westland Whirlwinds on fighter-bomber missions against targets in Occupied France and shipping off the French coast. He fired his guns once at enemy aircraft, and was not a fighter ace, nor a widely-known member of the RAAF. He is included here as representative of the majority: the young men who volunteered for the RAAF, flew the aircraft to which they were assigned and did their best. Today, they are remembered by their families and surviving squadron mates. Their names are listed on local and national memorials and they are mentioned in squadron records. It was their steady unspectacular dedication which contributed so much to winning the air campaigns.

Maxwell Tylney Cotton was born on 2 May, 1921 at Swansea on the east coast of Tasmania. He was the youngest of four children born to Mr and Mrs Tylney Cotton, of 'Kelvedon', a sheep and farming property some 20 miles (12km) south-west of the small town. In 1828 Francis Cotton from Hoxton, England, sailed to Tasmania and, after looking for suitable land, applied for a grant in the area of Great Swanport. Construction of 'Kelvedon' began in 1830, and at time of writing in 1990, the property remains in the family, being managed by Douglas Cotton, Max's brother. There were two sisters in the Cotton family, Jean and Mary. In 1935, Jean and her mother, Louisa, were killed when an aircraft taking off from the Swansea landing ground hit them. This happened in full view of the rest of the family and other onlookers.

Max Cotton was educated at the Friends' School in Hobart, went to work on 'Kelvedon' when he left school in 1936 and remained in that outdoor life until the war began. He volunteered for the RAAF but, like so many others, was told to wait until called. After being on the RAAF Reserve for seven months, he received what he described as 'an important looking letter from the Hobart Recruiting Centre'.

Max was to report to the office at 08.00 on the next Wednesday, ready to go to No. 1 Initial Training School at Somers, Victoria. Max was there at 07.30 and noticed first one, then another, young fellow standing short distances along the pavement. Of course, all were recruits, and by 08.00 there were 12 youthful volunteers ready to enlist as faithful servants of His Majesty King George VI.

Next morning, all 12 met again at Hobart railway station and went to Launceston, then to Melbourne by the ferry *Nairana*, which rolled and pitched its way across the Strait to the mainland; Max was quite ill. He had never been away from Tasmania, and only ever a passenger on small ferry steamers.

The draft arrived at Somers on 26 April 1941, were greeted with 'you'll be sorry' from the vastly more informed trainees already there on course, were issued with uniforms and other personal equipment and began life in the RAAF. That first night there were 12 quiet and shy young Tasmanians making the adjustment to service life, military discipline and the prospect of at least several years away from family and friends.

Elsewhere in the war, the Germans were advancing in Greece and North Africa, having broken through at Thermopylae and taken Halfaya Pass, pushed on to Athens next day, took Sollum the day after that and, by the end of the month, prepared to attack Tobruk and Crete.

Max had been accustomed to a hard physical life on the farm, and was the only one of the 12 without the Intermediate Certificate. He found the mental effort required was quite hard at first and he had to put in extra study to keep up with the others, remaining behind with the books while the other chaps went into Melbourne. He passed the allocation interview which saw the recruits divided into those for training as pilots, observers or wireless operator-air gunners and Max went on to pilot training. Max's sister Mary was nursing at Footscray, and he managed to meet her on the times he did get into Melbourne. As the one familiar face on the entire mainland, young Max was often loath to leave Mary and get on the night train back to Somers.

In the way of the armed forces, four of the original 12 Tasmanians returned to Tasmania for further training, while the others went elsewhere. On the way back, Max was again sea-sick. While at Somers, he had put on two stone (12.7kg), and returned to 'Kelvedon' looking fit and well. After five days' leave, most of which was spent going around the district visiting his relations, young Cotton reported to No. 7 Elementary Flying Training School at Western Junction. He also advanced from the lowest form of air force life, Aircraftman Class 2 (AC2) to Leading Aircraftman (LAC) and, more importantly, his pay increased from five shillings to nine shillings a day. These amounts are almost negligible by comparison with that in the 1990s, and quoting them as 25 cents and 45 cents respectively would give no true idea of the purchasing power of nine shillings in 1941.

It was 26 June 1941. Max's flying career was to last just eleven days short of two years. Since he had been in the RAAF, the Germans had captured Crete, the pocket battleship *Bismarck* had been sunk by the Royal Navy, British Commonwealth forces had invaded Syria and fought the Vichy French forces there, the USA had frozen all Axis assets, Germany had invaded the USSR and U-boats were inflicting heavy losses on shipping in the Battle of the Atlantic. Overall, the Axis powers were winning the war.

During his first flying lessons, Max 'had great difficulty in keeping the old Tiger Moth straight, and I thought—"Lordy!

I'll never be able to fly these damn things." However, after a few flights, I began to pick it up, little by little.' He found that landings were hardest, but was soon told that any landing you can walk away from is a good one. Max was impressed by, and admired, his instructor, Flying Officer Maloney. After seven hours dual, Maloney sent Max for a flight commander's check with Flight Lieutenant ('Silent') Knight, who climbed out, gave Max a short and clear talk on what he was to do, had him sign a little note book to signify that Max understood, and sent him off for the flight that all pilots remember.

With seven hours experience, Max was feeling confident, and recalled that he 'must have been busy feeling confident when I came in to land, instead of thinking of what I was doing.' As a result, the Tiger Moth ballooned down onto the grass, bounced back, but the now-alert Max caught it with full throttle, went round again, made a good landing and taxied back to the hangars, 'feeling as proud as punch.'

Flying training progressed, and at the end of the course, Fred Halcombe from Adelaide was first. Max was graded 'Slightly Above Average', but thought he did not do as well as he could have, coming 15th out of 30. Four of the course went elsewhere, and the other 26 were posted to Canada for further training.

The fledgling pilots were given eight days' leave before they left for overseas. Max spent his time at 'Kelvedon' or going round the district saying farewell. He spent half a day wandering over the hills and fields on the property, all of which he knew in fine detail after working there full time for the past four and a half years. He knew that he was saying goodbye to that part of Tasmania, and to Australia, for several years at least. Max was driven to Hobart by his father, and caught the train ferry to Melbourne. More farewells followed before the ship sailed, and it was only the young male's fear of looking foolish which allowed him to hold back the tears. 'I did not know how much my home and my people meant to me, till it came to leaving them,' he wrote.

But once at sea, the sadness was dissipated by meeting some of the other fellows from the flying course, and to make it all more enjoyable, Max crossed Bass Strait without any feeling of sickness. In Melbourne, Mary was in the middle of final nursing exams, but came to see Max depart on the train to Sydney. The carriages were crowded with pilots, observers and wireless operator-air gunner trainees from Victoria and South Australia, all on another stage of their journey to the distant war. After less than a day in Sydney, all the draft were to report at No. 2 Embarkation Depot at Bradfield Park, where they stayed for two weeks. There was plentiful leave, which was fun for a while but youthful impatience caught up and most wanted to move on again to Canada.

By the fortunes of war, the ship on which they were to sail to the USA was the luxury liner SS *Monterey*, and the 400 aircrew trainees travelled first class. As the 20,000 ton ship left on 18 September 1941, Max went down to his cabin 'with a lump in my throat which I could not swallow.' Other feelings soon intruded, as the Tasman Sea was rough and many of the passengers were sick, Max included. His cabin partner was Wally Rose, from Cloncurry, who was equally as sick. After welcome visits to Auckland, Suva, Samoa and Honolulu the ship arrived at Los Angeles. Shore leave had been allowed at the earlier ports, but not at Los Angeles. After a few hours, the ship left for San Francisco. The young Australians and New Zealanders were quite impressed with that city and with the bridges spanning the bay. They disembarked and went on by train, arriving in Vancouver on 9 October. The potential fighter pilots went on to Aylmer, Ontario, and those destined for bombers went to other Canadian bases. As a farmer, Max was greatly interested in

Formation flying in RCAF Harvards, Canada. (Douglas Cotton)

the land and crops passing by. On 12 October, the draft arrived at Aylmer and, as there had been no chance for a shower since Vancouver, first priority was luxuriating in the hot water.

Max found the North American Harvards at Aylmer 'good solid machines' and liked the 550-hp engines which 'would pull you up into the air like a rocket.' After 4½ hours dual, Flying Officer 'Happy' McLeod sent Max solo. After the simple Tiger Moth cockpit, Max found the larger number of instruments in the Harvard a challenge, and 'in doing a circuit I was busy all the time, adjusting RPM, or airspeed, or pulling up the wheels, or something.'

In a letter to his sister Mary, Max told her that he had gone solo after 4½ hours dual, but so far had done only 'circuits and bumps' and a few spins. Take-off and landing did not worry him, but he had problems with height while in the circuit, being busy with other matters and forgetting the altimeter.

'But they are real planes,' he added, 'and they will climb like tigers. (I don't mean Tiger Moths).'

The accommodation was far superior to that Max had experienced in Australia—as the climate demanded—and the hospitality of the Canadians added to the enjoyment of their time there. The Canadians were reputed to have the toughest training standards in the Commonwealth, and required much work from the trainees. Another aspect of the training, the war and the general adventure was that the Australians were meeting and getting to know well young men like themselves from other parts of their homeland. Max was not yet accustomed to the rapid succession of meeting new friends and then seeing them depart for another camp or course in the training schedule, and often regretted the breaking up of groups of friends.

In a harder school, he began to adjust to the steady trickle of fatalities as the combination of weather and inexperience resulted in fatal crashes. One of his good friends was Ralph Brockhoff, a big chap from Adelaide. With instructors, Max and Brock took off to practice formation flying, but ran into cloud, at which the instructor took over from Max and climbed through it. When they came into the clear, the other Harvard could not be seen, but after a time Max saw it diving into the cloud in an attitude which did not look right to him. Max and his instructor returned to Aylmer and later found Brock and the instructor were killed in the crash.

In December, Max wrote to Mary describing the local frozen pond, on which the boys skated. The natural surface 'put the Melbourne Glacierium in the shade for speed. I came one or two awful gutsers this evening, but some of the boys are pretty good at it.'

Training was broken by five days' leave over the New Year, and the great moment of Wings Parade came on 16 January 1942. By this stage, many of the boys had local girlfriends, and Max's came to watch the parade. It was a day he had been looking forward to, and Max felt he had achieved something as he marched out to be 'winged', but also he felt selfconscious in front of the crowd. The course remained for another month, as LAC pilots, for formation practice and night cross-country flights, before being awarded sergeant or commissioned rank.

On 3 February, Max described the winter surrounds, stating that they did not go outside any more than necessary as it had been −10 degrees Fahrenheit for the past three days. His skating had improved, but was nowhere near the standard of some of the locals, and he mentioned the usual sight of children of three or four 'tearing round on the ice.' The formation flights as ordered sometimes became too boring, and the Aussies would leave formation and indulge in aerobatics, or formation divebombing. Once they went down in formation to 30 feet (10m) from the ground, which all realised would result in punishment if they were caught low-flying. He had dived a Harvard at 300mph (480 kmph), and a couple of the others decided to beat that, but returned with damaged aircraft. One lost the cockpit canopy and the other had all the fuselage skinning peeled off from behind the engine to the rear seat.

Night navigation in Canada was not too hard, Max said, because the lights of the towns were visible long before the plane arrived overhead, and from farther away than in daylight.

On 14 February 1942 the course left Aylmer, each with US $90.00 and nine days' leave, with orders to report to Halifax, Nova Scotia, for the voyage to England. Max used his time and money to visit Toronto and New York, indulging in the usual tourist activities such as going to the top of the Empire State Building. Canadian hospitality was only surpassed by that of the Americans.

Finally, on 27 February, the SS *Orion* was boarded and the Atlantic crossing began.

By this time, the war news was worse than ever. Japan had entered the war and was claiming victory after victory in the Pacific; both the US and Royal Navies had been defeated; impregnable Singapore had surrendered; the Germans had advanced as far as Moscow before being halted by the winter conditions; U-boat sinkings in the Atlantic were numerous.

All aboard the ship were rostered for various duties, which included submarine watch. The food was far below the standard of the *Monterey* and there was no butter or jam — the bread had to be eaten dry. On 9 March the shores of Scotland near Glasgow came into view, and the airmen took it all in: convoys forming or dispersing, warships, barrage balloons, ship-building yards, flying boats swinging on the river.

Then they squeezed into English trains and crawled south through the late-winter countryside. As in Canada, Max noted the appearance of the farms. Patches of snow and ice remained, the trees were bare, but 'it was wonderful country. There were many freshly ploughed fields and rich soil. The farms were small, but neat and tidy, and prosperous-looking.' At Bournemouth, the draft was billeted in hotels, and for some days found time best taken up with sight-seeing, as there was no air force work for them. The Aussies were impressed by the 'business as usual' attiude of the local people. Overhead were reminders of why they were in England, as formations of aircraft set off for France, or returned. Of course, everyone wanted to be a Spitfire pilot. As the training system was unable to absorb them, plenty of leave was given, but the young men became somewhat impatient as day after day they watched fighters and bombers passing. Leave to London and other parts of the UK was enjoyable, but it was not why they had come all this way. On 21 April, Max was posted to 5 Advanced Flying Training

Unit, Tern Hill, Shropshire, with a number of other boys from Aylmer.

Mail deliveries were slow in wartime, with little airmail. On 28 April, Max acknowledged receipt a few days before of a letter from Mary dated 20 December. He commiserated with her on the 'flat' Christmas of 1942, under wartime conditions. He gave a few brief comments on the Miles Master aircraft he was flying and described the contents of an issue from the Australian Comforts Fund (ACF), which included personal items and, the thing he liked best, a packet of ginger nuts, adding, 'I don't know why, but I'm always hungry in this country. Just plain gutsy, I guess.' The letter, and many following to his family, was written on Australian Comforts Fund (ACF) letter-paper.

The Miles Masters they flew at Tern Hill were found inferior to the Harvards when doing aerobatics but easier to land. On 2 May, Max celebrated his 21st birthday, enlivened by a cake provided by Dick Lloyd from Melbourne and a visit to the local pub, as well as the fortuitous arrival of nine letters, including five from home. By the time they left Tern Hill, spring was quite evident, with high grass, bright green leaves on the trees, and to Max, 'no man who has lived in the country all his life could help being impressed by such a sight.'

After soloing on Hurricanes, Max was posted to 59 Operational Training Unit (OTU), Crosby-on-Eden, Carlisle, Cumberland. Once again, he lost a friend in the posting system when Fred Halcombe went to another unit. They had been together since Somers camp in Victoria. On 12 May Max reported to 59 OTU.

Now they learned to fly Hurricanes and, even if the aircraft were old ones from squadrons, they were fast enough for Max and the others—at first. He wound one up to 320mph (515kmph) before he had been at 59 OTU for a week.

Despite constant warnings and the best efforts of the staff, there were some fatal crashes and mid-air collisions. Like many others, Max used to do a little aerial touring, admiring the numerous castles and the nearby Lakes District. One day he was cruising at 500 feet over a lake and suddenly realised that the mountain sides were closing in. There was not room to turn and he only scraped over the mountain at the head of the lake by going to full throttle and climbing 'like a homesick angel'.

In a letter to Mary, he said he had 'the Hurricanes more or less under control now, and can throw them around the sky. Doing aerobatics in them is quite a pleasure. I had never tried upward rolls until the other day, and they go quite well, but you need at least 350mph before you start.'

High spirits! A Hurricane 'beat-up' at 59 OTU, 1942. Note the men ducking in foreground. (Douglas Cotton)

He mentioned that the course began with 44 students, but only 41 remained; a New Zealander had hit a mountain and been killed that day. 'We have some bloody awful weather here, and it closes in very quickly. I'm getting used to my pals being bumped off, now, and perhaps it's just as well. There hasn't been any more news of our postings, but as far as we know, it's the Middle East.' Max, like thousands of others, was to go to anywhere but the rumoured destination.

In the air-firing phase, Max scored fairly well, on one flight destroying the target drogue. There was also air-sea firing, the oxygen chamber and altitude tests, cross-country navigation and practice dog-fighting. The end of course frivolities at nearby Longtown escalated into a wild free-for-all in which the Sergeants' Mess windows and furniture were damaged, the floor was awash with beer, fire extinguishers were emptied on all and sundry, including those who tried to go to bed, and one man rode a horse into the building, demanding to have his photograph taken.

Next morning, all the revellers had to go to Crosby and find out where they were posted. Max, Frank Hicks from Tasmania and Adrian Browne from Sydney were to report to 87 Squadron at Charmy Down, Somerset. Again, friends made on course were farewelled. Most were going on leave to London, but Max decided to go north, to Edinburgh. After a week in Scotland, he thought it the most beautiful part of the UK. On the way south again, Max was repelled by the bomb damage in Bristol, where entire streets had been demolished. The contrast with the unspoiled north was such that he did not stay. 'It looked so forlorn, and the faces of the few people I saw were so glum that I didn't stop to see any more, but caught the next train to Bath.'

The luggage of Hicks and Browne were at the railway station, and Max took it all by taxi to Charmy Down airfield. Disappointment awaited him. 'When I walked in the gate and found Hurricane nightfighters, instead of Spitfires as I had expected, I could have screamed.' To make matters a little easier, there were already three Aussies on the squadron, but it was nonetheless a severe disappointment to the budding fighter pilots. He added more in a letter to Mary, commenting that he was expecting a Spitfire squadron and when he saw 'dirty black Hurricane nightfighters, I could have wept.'

However, the new boys were told that if they did not really like the single-seat night-fighters after a month or so, the commanding officer would arrange for them to be posted to a dayfighter squadron. After a few night flights for experience, they began flying interceptions, but this was fruitless. Responding to radio directions from a controller, the fighter would be guided to the vicinity of a German bomber which it would try to shoot down with cannon or machineguns. There were few successes and, after only two weeks, when the new arrivals were asked if two of them would volunteer to go to 263 Squadron, flying Whirlwind fighters from south Wales, Max and Frank Hicks took the opportunity.

Max was not keen to fly twin-engined fighters, but reasoned that there would be more action than with the nightfighters, and had persuaded Hicks to go with him. Another factor influencing Max was the knowledge that soon 87 Squadron was to convert to the Beaufighter, and undergo the intense training necessary for landing by radio at night, the Beam Approach system. He had done it in a Link trainer, and been put off by the bellowing of the instructor. He thought it would be worse in a real aircraft.

On 31 July the two Tasmanians went by train to Angle, South Wales, to join 263 Squadron. Again, Max was appreciative of the natural beauty of the locale, but at Angle there were a few extra aspects to keep his attention. The airfield was on a peninsular, with the sea on one side and Milford

Haven Harbour on the other, which meant steep cliffs all along the seashore, and the runways ended close to the cliff tops—not conducive to overshooting on landing. They had never flown a twin-engined aircraft, and there was no dual-seat version of the Whirlwind. Before being sent off in a Whirlwind, the new boys were given 45 minutes in an

Westland Whirlwinds in formation. (Douglas Cotton)

Oxford. This was a docile machine, which Max termed 'sloppy' on the controls by comparison with those previously flown. After lunch, they were shown the Whirlwind cockpit, and had it explained to them. 'The cockpit had considerably more dials in it than the Hurricane or Master,' said Max, 'and undercarriage and flap levers were in different positions to the Hurricane, and took some getting used to.' Once the cockpit layout was familiar to them they were allowed to take the Whirlwind up. 'It was quite a pleasant aircraft to fly,' wrote Max, 'you didn't notice the bumps because she was so heavy.' The double number of airscrew and pitch levers took some getting used to, and it seemed strange having an engine on each side rather than one in front, but Max adjusted rapidly. 'All went fine till it came to landing. That's when the fun started. They are supposed to be one of the hardest aircraft in the Service to land. They have to be flown in at 120mph (200kmph) and there are very few planes with as high a landing speed as that. We were advised to do wheel landings and allow the tail to go down after she had run along for a while.'

Max's first approach was too low, and he went round again after seeing the red light. Next time he stalled too high, 'and she hit the ground with an awful wallop, so I slammed on the motors and went round again. Next time I made a reasonably good landing.' After a talk to the flight commander, Max went up again to familiarise himself with the local area. On return, he made 'a bloody awful landing. They certainly make the Whirlwind undercarriage strong, or it would have folded up on me.' Then, when he practised his landings, the port throttle jammed and he came in with one engine idling and one roaring at full throttle. 'She reared and plunged down the runway, and if I had not switched the engines off I would have ended in the sea. By that time, my hair was just about turning grey, though any landing you can walk away from is supposed to be a good one.' Max had a thorough introduction to the Westland Whirlwind.

Frank Hicks, known as 'Butch', and Max were the only two Aussies on 263 Squadron, which was made up of English, Welsh, Scots, Irish, South African, Canadian, British Guianan and Singaporean, reflecting the unique wartime mixture of the RAF. The squadron was commanded by Squadron Leader R. Woodward DFC, who was quite experienced on Whirlwinds. What this variety of men had in common was a large number of hours on Whirlwinds, from 200 to 500 hours, and the two Aussies felt very much the raw beginners.

They could not be allowed on operations until masters of their aircraft, and much flying practice followed. Gradually the two became more expert, and tried pushing the Whirlwind to its limits. Max reached 500mph (800kmph) in a dive once and was content with that. He found that by putting down a little flap the plane had a reduced turning circle, but he realised that it was not capable of turning with current single-engined fighters and was not as manoeuvrable. 'In fact,' said Max, 'I did not quite see what was the purpose of the Whirlwind, unless it was designed as a bomber-interceptor.' The Whirlwind had been flying operationally since 1940, and engaged Messerschmitt Bf109s on several occasions, but was not really a match for them. Just over a week before he and Hicks arrived, 12 Whirlwinds with Spitfire escort had engaged 109s who shot down two Whirlwinds. Apart from flying, there was not much of interest around Angle. German bombers came over once, attacking ships in the harbour, but lost two Ju88s to the Beaufighters. On 19 August, the squadron moved to Colerne, and simultaneously Frank Hicks was posted temporarily to Boscombe Down for a period of flying test aircraft. He was the last Aussie of the group which had gathered at the Hobart Recruiting Office with Max.

As a junior pilot on a fighter squadron, fully occupied with

Max Cotton DFC

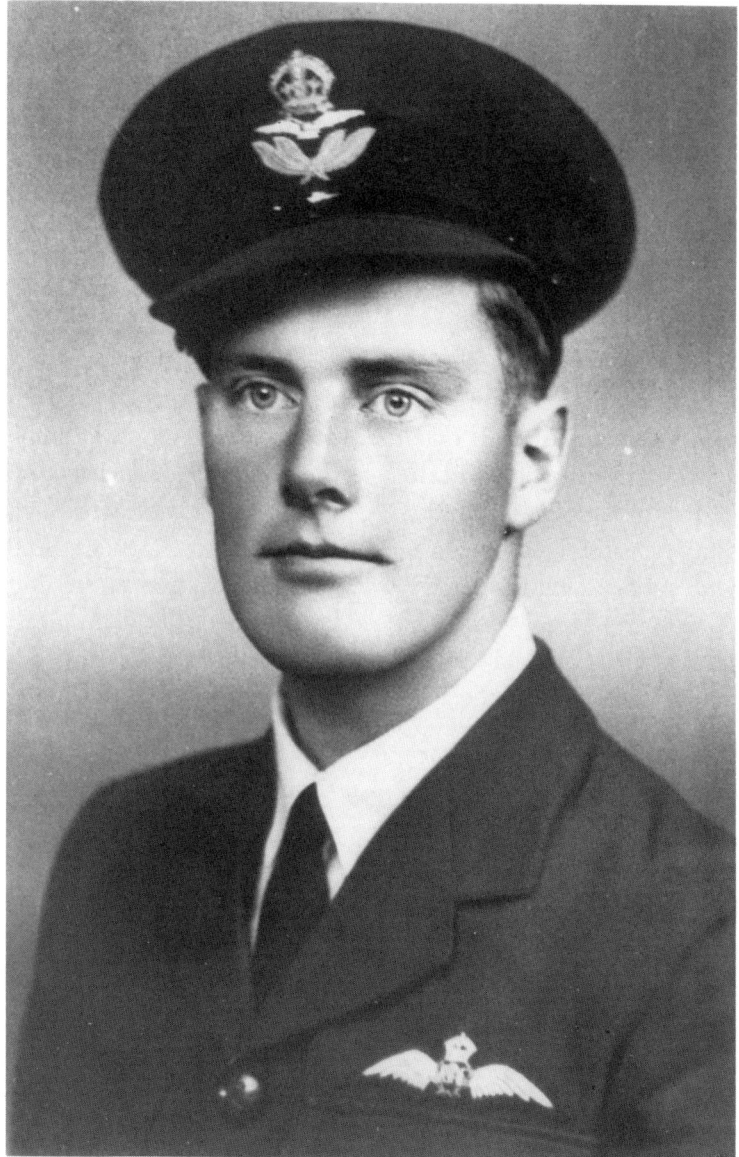

Pilot Officer Max Cotton DFC RAAF, 263 Squadron RAF Westland Whirlwinds. (Douglas Cotton)

mastering his aircraft, Max seemed unaware of the larger war horizons.

After the terrible first Russian winter, the Germans renewed their offensives in the East, and achieved more great victories over the Soviets; Malta was isolated in the Mediterranean and fierce convoy battles were fought to bring a few ships to the island; Rommel had inflicted more defeats on the Commonwealth forces in North Africa, had taken Tobruk and was only 60 miles (100km) from Alexandria and the Suez Canal; U-boats in the Atlantic were sinking as many ships as ever; battles in the Pacific were fought between aircraft carrier strike forces, and the US claimed victory in the Coral Sea and at Midway; Burma had fallen to the Japanese, who also had landed in New Guinea and were advancing on Port Moresby; Japanese submarines had penetrated Sydney Harbour and were active along the southern and eastern coasts of Australia; Darwin had been bombed and raids continued. On 19 August, the British-Canadian force, landed in front of the defences of Dieppe, was annihilated and the air battles overhead were the most intense of the year.

At Colerne, the Whirlwinds were fitted with bomb-racks, while the pilots waited and watched the other resident twin-engined squadrons operating their Beaufighters and Mosquitos from the base. On 23 August, Flight Lieutenant C. Rudland flew a Whirlwind with two 500-pound bombs under the wings, and at sea-level reached a maximum speed of 270mph (430kmph). The usual maximum speed was 300mph (480kmph). After flying some 20 hours on Whirlwinds, Max was deemed 'operational', which meant he took his turn at standing readiness. He explained that while this sounded exciting, it was actually boring, when those on duty sat about for hours; nothing happened. Firing practice with the 20mm cannon in the nose of the Whirlwind revealed another aspect of the aircraft. Max found it 'used to buck so much when you pressed the trigger, made a terrific din and clouds of smoke. I found it much harder to shoot accurately with cannon than with machineguns.' The four cannon were mounted in the extreme nose, but literally just in front of the pilot's feet.

While at Colerne, Max spent a day of leave helping an old farmer with the harvest. There was a shortage of labour, and Max welcomed the chance, but regretted it next day when stiff muscles made it obvious that it was a long time since he had done any farm work. The rest of his leave was spent in London and Maidstone, visiting Doctor Warren, an uncle of his father's second wife, Frances. There were four sons, all doctors, and two daughters. He saw more of England, and thought Kent the prettiest county he had visited. 'The farms were very green, and neat and tidy. There were many interesting old cottages with thatched rooves and oaken beams.'

Back at Colerne, more new pilots had arrived so Max was not the junior of the squadron, and the new fellows just out of OTU grappled with the mysteries of twin-engined fighters. On 7 September 263 Squadron moved to Warmwell, in Dorset. This was a grass airfield, the first Max had used since elementary training in Tasmania. The clumps of grass meant that bouncy landings in Whirlwinds became more common. All the accommodation was taken by other squadrons, and 263 had to store its parachutes in a tent. Accidents were frequent at Warmwell and once Max heard the driver of a crash-truck, arriving at a wrecked Spitfire, remark to his mate that he wished the pilots would go somewhere else to kill themselves.

However, the bomb-racks meant a definite purpose in life, and bombing practice began in earnest. There were no bomb-sights, so actual dropping was by the pilot when he felt the aircraft was in the correct place. The target was a triangle with sides of 15 feet (5m) anchored offshore. Max described the procedure as: 'to fly low over the water about 5 to 10 feet up (1.6 to 3 m) at 250mph (400kmph), and release the bomb just before the target disappeared under the nose of the aircraft. You must press the bomb release button at just the right moment.' After repeated runs over the target, the bombing error was reduced, from 40 to 20 feet (12m to 6m), 'then I amazed myself and everyone else by getting six direct hits out of eight, and an error of three yards (metres). After that, I'd "got the game taped" and could get four or five direct hits every time.' However, young Sergeant-pilots were not senior or experienced enough to go on the first Whirlwind bombing strikes.

On 8 September, 263 Squadron was the only one available for shipping attacks off the coast of France, as 174 and 175 Squadrons had been stood down. That afternoon, the first strike was flown by Whirlwinds. Four armed with two 250-pound bombs each took off, escorted by the Exeter Wing of Spitfires, but no ships were found. However, next day, again with Spitfire escort, four armed trawlers steaming west between Cap de la Hague and Alderney island were attacked, two ships being sunk. The squadron was amused at press descriptions of the attack, particularly references to Whirlwinds attacking at 400mph (640kmph). The Westland company sent a letter of congratulation to the squadron for its achievement in increasing top speed by carrying two 500-pound bombs.

Things did not always work out well, and on 3 October four Whirlwinds flew to attack a 3,000 ton motor vessel in Alderney Harbour. Only one of the Spitfires delegated to attack the flak positions crossed the breakwater with the Whirlwinds, the others having become late. Three Whirlwinds attacked the target and the fourth bombed another ship, but because they were going at top speed across the harbour, no one saw any results of the attack. Two aircraft were hit and Squadron Leader Woodward returned to base on one engine.

Despite the commencement of operational strikes, practice continued. Once bombing practice was over for the day, the pilots went on leave, going to dances, films or pubs. High spirits led to 'souveniring' of things, including a 10 foot (3 m) dinghy which was relaunched in the camp water supply.

Finally, with 50 hours on Whirlwinds, Max was included on the attack sorties. His first mission was on 10 October, as No. 2 to Pilot Officer Jimmy Coyne RCAF, looking for shipping around Alderney and the Cherbourg Peninsular. Nothing was seen, and they returned to base, Max dutifully weaving so as to look behind and see fighters approaching. Even though they were at very low level, German radar picked them up, and the distinctive buzzing in their earphones warned them of enemy radar capability. Another sortie was flown on 13 October, again with no contact. After this, a reconnaissance aircraft looked for ships, then flew back and alerted the strike force.

Max described the flights in a letter, grumbling about the useless carriage of bombs out and back, and wrote of the party in Weymouth to celebrate the award of the DFM to one of the squadron. 'We celebrated at the rate of twelve pints an hour, and soon became horribly drunk. On the way home, we grabbed a large sign from outside a restaurant and put it in the back of our truck. It now stands outside our Flight, with a plane painted on it. I felt very off-colour for a couple of days afterwards. We have a booze-up two or three times a week to relieve the monotony, because it does get rather boring at times, messing around, doing a couple of hours flying each day and nothing the rest of the time.'

Max went to London for a week on 29 October, and arranged to meet Bruce McKenzie, one of the Hobart group, now on Typhoons in Norfolk. They spent much time at Australia House, reading magazines from home, eating, and meeting other Aussies—including friends from the Aylmer course who turned up one day. Comparing notes, Max was sorry to learn of the many who had been killed. The togetherness at Australia House was remarkable and very pleasing to Max, missing as he did the company of other Aussies. He remarked that one never knew who would be met there, anyone from an AC2 to an Air Commodore. Two more of the Hobart group arrived after training in Rhodesia (Zimbabwe), with commissions, but had not yet been to an OTU. Meanwhile, Frank Hicks returned to 263 Squadron, with a fund of anecdotes about his experiences at Boscombe Down, including finding himself in the London balloon barrage with

Westland Whirlwind P7062, 263 Squadron RAF. This aircraft was normally flown by Sergeant Max Cotton RAAF, but was destroyed on 17 February, 1943 when flown by Flight Sergeant 'Butch' Hicks RAAF, who hit a tree on low flying exercise. Sergeant Cotton flew 34 training and operations sorties in this aircraft. (Douglas Cotton)

only 15 gallons (67l) of fuel in the tanks. The Commonwealth was further represented by the arrival of two Hindu pilots, Pilot Officer Samant and Sergeant Thyagarajan, known as 'Bill' or 'Tiger'.

On 7 November, Max flew his third sortie. For much of the time, he had used Whirlwind P7062, and flew it on this attack also. Led by Pilot Officer DR Gill, four aircraft set off, splitting into two pairs at the French coast, to attack railway lines and trains. Again Max was No. 2 to Jimmy Coyne. The weather was foul, very cloudy, poor visibility and teeming rain. 'I was so busy concentrating on keeping with my leader that I did not have much chance to look for targets. We were at tree-top height.' When Max did look around for a target, it was at the moment that Coyne turned towards him, and they almost collided. Suddenly there was a railway station there before them. They dropped their bombs and Coyne fired his cannon at a train, then they were past and gone. Looking back, Max was pleased to see the explosions in the goods yards. At last, after 15 months training, he had struck a blow. Max noted that he was only frightened before take-off, but once in the air 'I was OK. When I got over to France and started blowing things to bits, I wondered why in hell I ever got frightened.'

The other pair of Whirlwinds was also struggling through the weather. Sergeant Abrams saw Gill's bombs explode on or near the track, but then lost sight of him in the rain. Gill did not return, and it was believed he had flown into the ground or an object. Cloud base was only 250 feet, and the rain was so heavy that relatively clear vision was possible only through the side cockpit panels.

Another strike was flown on 19 November. This was the first of all-Whirlwind strikes of a series termed 'Roadstead' operations, this one being Roadstead 40. Different types of offensive operations had codenames but, for simplicity, they will be termed here as strikes or sweeps. However, definitions may be of interest:

Circus was a large scale combined fighter and bomber operation designed to bring enemy fighters to action so consequently had many fighters, few bombers and a short radius of action;
Ramrod was a bomber force escorted by fighters, but the intention was to destroy the target and was to the extreme fighter range;
Rhubarb was a small harassing operation by fighters or fighter-bombers, intended to destroy low-flying enemy or ground targets;
Roadstead was an attack on ships at sea or in harbour, with fighters escorting the attacking force, whatever the aircraft type;
Rodeo was an offensive fighter sweep without bombers;
Rangers were deep penetration sorties into the enemy rear, intended to destroy the types of aircraft found there.

On this Roadstead, Max flew as part of the anti-flak element, in A Flight, as No. 2 to Squadron Leader Woodward, with Jimmy Coyne RCAF and Harvey from Eire as the other pair. The ships had taken refuge in harbour and the aircraft returned to base. Another similar sweep was flown next day, but Max was not with them; there was no contact.

On 7 December, Squadron Leader Woodward and Warrant Officer McPhail were killed on a shipping strike. Eight Whirlwinds were attacking a convoy south-west of Jersey, through intense flak, when Woodward was hit and went into the sea some 400 yards (metres) from the ships; McPhail flew into a burst of heavy flak, and went straight into the sea. The squadron command went to Flight Lieutenant Geoff Warnes, later described by Jimmy Coyne as 'the finest leader of men I have ever known.' Warnes was a burly Yorkshireman, who had played rugby for the county pre-war. He may have been the only operational pilot in the Second World War to have worn contact lenses.

More two-aircraft flights to France were made, with railways as target, but weather was not good enough and there were no actual attacks.

Squadron Leader Geoff Warnes, Flight Commander and Squadron Commander, 263 Squadron RAF 1942-43 and 1944. (Douglas Cotton)

Max was amused to read a letter from Mary, who had drunk a few whiskies to cure a cold. He replied that he 'got soaked once a week, regularly, but am afraid I haven't always got a cold to cure. Smoke like a chimney, too, so I guess I have gone to the dogs. It relieves the monotony a bit, to go into town and get merry.' He had received a pair of shoes sent from home four months previously, and also a cake. 'The boys hovered round me like a lot of vultures for the raisins and walnuts on top. But I haven't cut a hole in the cake yet. A week ago, I got a parcel of dried fruit from Frances and Dad, which was very welcome. They don't grow much fruit in this country.'

The fast and powerful FW190 had been used by the Luftwaffe for nuisance raids, in RAF and cricketing parlance, 'tip-and-run' raids. The Germans would fly in at low level to the English coast, bomb and strafe whatever presented itself as a target and be away before the defences could react. Reaction time measured in seconds was necessary to intercept these attackers, and the only fighter in the RAF capable of catching the 190 was the Hawker Typhoon, which had suffered airframe and engine troubles early in squadron service. The procedure adopted was for two Typhoons, pilots strapped in, to be positioned at the end of the runway, with groundcrew and starter-motors ready. A flare would be fired from station HQ and in a few seconds the Typhoons would be taking-off to receive directions when airborne. The pilots had to be changed every hour or so after sitting strapped into the cramped cockpit. The Typhoon Squadron nominated for duty on 14 December was 266 (Rhodesia) Squadron, who wanted to celebrate their National Day on 14 December, so arranged for 263 Squadron to take over the readiness duty.

Jimmy Coyne and Max were at the end of the runway, well aware that the FW190 would be a handful, even if they could catch one. Whirlwinds had shot down German bombers and a few Messerschmitt Bf109s, but had not fought the 190. Both pilots were aware of the formidable reputation of the Focke Wulf and of the trouble even experienced Spitfire pilots had in coping with it.

They were sent off to intercept an attack on Poole, and directed onto several courses looking for the intruders, but with nothing seen, were turned back to England as the radar controller had lost the Germans. By this time, the Whirlwinds were low on fuel. Jimmy Coyne saw the Germans, and turned towards them, but the first Max knew was when he saw tracer streaming past, very close, from the sun. The FW190s were attacking. They were only 15 miles (24km) from France, with cloud over 5,000 feet above, and only 15 gallons (68l) of fuel each.

But Max realised afterwards that none of these things were considered—he just pushed the throttles forward, pushed the airscrew pitch lever into 'fully fine', dropped flaps to turn tighter and reefed the Whirlwind into a vertical bank, then turned 'for dear life.' Reflector sight was on, and gun button was already on 'Fire'. 'I was far too busy to be frightened. Things went round in vicious circles for a while, and I got a couple of shots at the German No.2. The first time I fired the Jerry was going over the top of me, and I was pointing straight up in the air. The four cannon made a terrific din, and my airspeed dropped off so much the plane stalled and almost went into a spin.

'I soon recovered and got to work again. Next time I fired, I noticed Jimmy shooting at the same time, and clouds of smoke pouring from one of the German planes. They must have got frightened, because they climbed out of our reach and beat it for home. Glancing at my fuel gauges, I decided beating it for home was a good idea too, because there was only 10 gallons (45l) in each tank and we had 50 miles (80km) to go back to England.'

Coyne had managed to get onto the tail of a German twice, and fired long bursts, scoring hits. Later, it was found the Germans were reluctant to get into a hard turning match at low level, as the 190 had a tendency to flick and spin without warning—often with disastrous results. At about 130mph (210kmph) the Focke-Wulf could stall suddenly and without warning, with the port wing dropping violently, and if in a turn, the aircraft could flick the opposite way and spin. Stall on landing with wheels and flaps down was quite different and acceptable, but fighting in turns was to be avoided.

Coyne was concerned about fuel also, and twice radioed that he did not think he could make it, but a tail wind assisted them, and with throttles right back, mixture as weak as possible to conserve fuel, the Whirlwinds returned safely, first landing at Hurn, then going on to Warmwell. In fact, as Coyne taxied to the Watch Office at Hurn, one engine ran out of fuel. A design fault in the Whirlwind was that fuel could not be transferred from one tank to the other. All the while Max was cursing his bad shooting and regretting the escaped Focke-Wulf. After a while, he considered they had both been lucky that the Germans were such poor pilots, as they had the advantage and attacked first. Congratulations came in, as they had shown the Whirlwind could deal with the FW190, and with the extra experience and confidence from his first combat, Max was 'itching to mix it with the Luftwaffe again, FW190s or anything else!'.

Earlier, with Pat Yates from Ireland, he had been for the preliminary interview with the Station Commander, as part of the commissioning procedure for Sergeant pilots. This was followed by the interview with the AOC of 10 Group, and both young men felt nervous in the august presence. 'However,' said Max, 'he was a very decent chap, Air Vice Marshal Dickens by name, and he didn't ask any particularly awkward questions.' Both sergeants were confident they had passed, but Max was aware that his commission would take longer, having to go through RAAF HQ as well as RAF Air Ministry.

Four days' leave over Christmas was spent with the Warrens at Maidstone, 'a grand time', though Max was disappointed in not seeing a white English Christmas. It was cold enough for an Australian, though. Dinner was an elaborate affair, with turkey, ham, pork, pudding and other delicacies. There were many parties and a dance on Boxing Night which was particularly enjoyable, as the women wore long dresses, a welcome change after the depressing sight everywhere in London of uniforms. Max had studied art and sketching at school, so enjoyed a morning sketching with one of the Warren grand-daughters. Then it was back to the squadron, and the bone-chilling weather at Warmwell.

By the end of 1942, Max had flown six different types of aircraft, ranging from Tiger Moth to Whirlwind, for a total of 273 hours, including his 50 hours on Tiger Moths. Of these, 108 hours were on Whirlwinds, with 16 hours on operations.

In the larger war situation, the Germans were suffering their second winter in the East, and the Sixth Army was surrounded at Stalingrad. General Montgomery had defeated the Axis forces at El Alamein and was steadily advancing west along the coast of North Africa, while an Anglo-American force had landed in Morocco in November and was pushing east. In the Pacific, Australian forces had defeated the Japanese at Milne Bay and on the Kokoda Trail, and were engaged in annihilating the remaining enemy in Papua. In the Solomons, after some ferocious land and sea battles, the Americans were prevailing over the Japanese. At last the balance seemed to be tilting to the Allies.

At this time, the German day fighter force in France

consisted of some 300 FW190s, belonging mainly to JG2 *Richtofen* and JG 26 *Schlageter*. In addition, another 140-odd were available from the Reich defences. These fighters were husbanded for attacks on daylight medium and heavy bomber formations which did most damage, but there was always the possibility they would engage the small anti-shipping forces over the coastline and Channel.

On 2 January 1943, Max was one of four pilots detached to Predannack, in Cornwall, tasked against German minesweepers operating near Brest, which enjoyed formidable defences and which the Whirlwind pilots had no wish to test. Of Predannack, Max wrote that he had 'met some windy cold places in my day, but nothing to compare with that joint. It rained hard most of the time, there was a south-westerly gale blowing up from the Bay of Biscay, which reached 60mph (100kmph). We had no extra clothing, so shivered most of the time.' To make the detachment feel more confident about operating over the long stretch of water to Brest, groundcrews at Predannack knew little about their fighter. 'Whirlwinds?' one of them said, 'never heard of them.' The four pilots did their own refuelling, re-arming and daily inspections. Worst of all, they had to leave a warm bed at daybreak to start the engines.

On 10 January, they set off with a Spitfire escort from 19 and 130 Squadrons, but no minesweepers were found. Max remarked it was comforting to look up and see the Spits around them, but added that no one was fool enough to go into Brest harbour looking for ships to attack. They returned to freezing Predannack, 'brassed off' in the vernacular of the day, wishing for a recall to Warmwell. After a fortnight, their sole uniform was a little grubby, and clean clothes were regarded as most desirable. Then four relief pilots arrived in an Oxford, and the exuberant quartet took turns flying it back to Warmwell, putting it through all sorts of gyrations.

However, discretion applied, and one who wanted to loop the venerable machine was firmly restrained for fear the wings would come off.

On 15 January, Squadron Leader Warnes, Flight Lieutenant Blackshaw and Flying Officer Eddie Brearley flew the first 263 Squadron night operation. Nothing was found around the Channel Islands, but Brearley bombed a train in France. The remainder of January and early part of February was taken up with shipping reconnaissances, attacks on rail targets, practice flying and bombing and gunnery practice. Other events filled the days, and on 9 February Sergeant John Macaulay in P6991 had an engine cut on take-off, resulting in a 'prang.' On the 12th, Dai Williams, a Welchman, went into the sea, and was believed to have been captured. Williams and Flying Officer Harvey were attacking trains in France, flak hit Williams' fighter, he turned for home, was unable to climb and ditched four miles (6.4km) off the French coast. Both propellors were ripped off and the Whirlwind bounced across the waves, but Williams got out, inflated his dinghy, climbed in and waved to Harvey. From Warmwell, Harvey returned to the location in an Air Sea Rescue flying-boat; there was no sign of Williams. As the wind was blowing to shore, it was presumed he had been taken there.

On 15 February, a Rhubarb operation was flown, but cancelled as the weather was unsuitable for sweeping over enemy territory.

Frank Hicks and three other pilots were flying in an Army co-operation exercise on 19 February, making low-level mock attacks on the trucks. Hicks was flying Max's usual aircraft, P7062. Max presumed that Hicks must have been concentrating on the trucks instead of what was in front of him, as his port wing hit a tree, the Whirlwind flicked onto its back, and Hicks tried to climb by pushing the stick forward, but damage to the wing was too great and he hit the ground some 400 yards (metres) further on. Pieces of the Whirlwind were scattered over a large area. Hicks was buried at Wroughton, in Wiltshire. For Max, this was a tremendous blow. Frank (Butch) Hicks had been with him since the first morning at the Hobart Recruiting Office, and in addition was married with two children in Tasmania.

On 20 February, Max and A Flight moved to the small airfield at Harrowbeer, and B Flight went to Fairwood Common, near Swansea. On 26 February, with a few belongings, they moved to Ibsley, just north of Bournemouth. Six Whirlwinds, with Spitfire escort, were to dive-bomb Maupertus airfield, on the Cherbourg peninsular. With the benefit of hindsight, Jimmy Coyne realised that the Whirlwinds were bait, to bring up the Luftwaffe fighters so the Spitfires could engage them. The force took off at 15.00, and on the climb to 15,000 feet, Max was a little pleased to note that even with bombs under the wings, the Whirlwinds could outclimb the Spitfires. At the briefing, the approach to the target had been given as from the east, but the winds were different, and the six Whirlwinds passed directly over Cherbourg itself. No one noticed this error until 'the anti-aircraft boys at Cherbourg showed what they could do.' Tracer flashed up, exploding shells patterned the sky around them, and shock waves rocked the Whirlwinds; one Spitfire was damaged, and it returned safely to England.

One after the other, the Whirlwinds rolled into their dives, reaching 400mph (640kmph), pulled out at 5,000 feet and released their bombs on the runways and buildings. Continuing in a more gentle dive, the Whirlwinds sped out to sea, going down to sea-level and home again. The Luftwaffe remained on the ground. The operation was repeated next afternoon, but the formation avoided Cherbourg. There was a lot of flak from the airfield defences, and the Whirlwinds went down through it, scoring further hits. Once again, no enemy fighters were seen. It was too good to last, of course.

Max Cotton DFC

On 28 February, B Flight flew down from Fairwood Common, and attacked a few minutes before A Flight. Max was last in the A Flight formation, and went down lower than he should have, to make sure his bombs hit the selected target. When he recovered from the dive, he was about 800 yards (metres) behind the other Whirlwinds — and many Luftwaffe pilots had become aces by picking off stragglers.

Max pushed the throttles fully forward in an effort to catch up with the flight, and noticed a Spitfire breaking overhead to get behind him; he thought no more of it, watching the distant Whirlwinds. The Spitfire was 'Z' from 130 (Punjab) Squadron, flown by a Free French pilot, Flying Officer Jacques 'Jaco' Andrieux. When diving with the Whirlwinds, 'Jaco' had seen two FW190s off to the left and above, reported them to the Spitfire leader, Squadron Leader Wright, but as the attack progressed the 190s were lost to sight. As the formation left the target area, one of the Spitfire section leaders called that a Whirlwind was straggling. Andrieux looked back and saw two Whirlwinds lagging, and also two FW190s diving on them.

'Jaco' broke left, towards the 190s, who were 1200 yards (metres) behind Max and obviously preparing to attack in close formation. He turned directly onto the 190s, intending to fly between them and split up their formation. Now the Spitfire leader turned the rest of the formation to the right, and Andrieux lost sight of them, but concentrated on the Germans, who saw him and broke hard left and away. 'Jaco' turned right and swung in behind the 190s, estimated range as 600-800 yards (metres), and realised that he was not catching them, despite 320mph (510kmph) on the airspeed indicator. He aimed at the starboard 190 and fired, thought the rounds passed below the German, so lifted his nose and fired again, saw black smoke pour from the port side of the 190's engine and fired once more. Heavier smoke streamed from underneath the cowling and the 190 slowed abruptly, then turned

Flying Officer Jacques Andrieux, the Free French Spitfire pilot who shot a FW190 off the tail of Max Cotton's Whirlwind. (J. Andrieux)

left for the coast, some 5 miles (8km) away. The other 190 went up steeply into the sun so 'Jaco' turned his attention to this potential danger, then decided he had best get back to England. He had fired 100 rounds of 20mm cannon and 220 rounds of .303-inch machinegun ammunition from Spitfire 'Muzaffargarh', and landed 20 minutes after the rest of the Spitfires and Whirlwinds.

Sergeant Bid, a Belgian, was flying as Blue 4. He also saw the straggling Whirlwinds, and two fighters behind them, but thought two Spitfires had been detached to close up on the Whirlwinds. He remained with his section, but turned again when he heard 'Jaco' call on the radio, thinking a Spitfire was in trouble. He saw a large splash in the sea, and a cloud of black smoke. After landing, he asked if it was a friendly or an enemy aircraft which had gone into the sea, but no one knew until 'Jaco' arrived. This independent report of Andrieux's success confirmed the victory for him.

Max entered the event in his flying log book, to which Squadron Leader Geoff Warnes added, 'Moral—Don't Straggle.' Max admitted learning two things from the incident: to keep up with the flight and to keep a sharp lookout behind, 'even with every bloody Spitfire in the RAF escorting.' He realised that if Andrieux had not shot accurately, the best he could have looked forward to was a day or so in the Channel after being shot down; alternatives were all worse.

Max's commission came through, and he was given leave to go to London to be properly kitted out, which included uniforms from Saville Row. Jimmy Coyne recalled that the entire squadron was pleased at Max's promotion.

On 12 March, the squadron returned to Warmwell, now regarded as 'civilisation', and next day Max took over P7108 as his aircraft. But on 14 March, they flew again to Ibsley and attacked Maupertus once more. Max flew no other operations in March, though there were detachments to Bolt Head and Predannack, and more practice flying, with camera-gun attacks.

On 4 April a sweep was flown looking for E-boats, but none were found. E-boat was the term used for enemy torpedo boats, and covered several types of small, fast and well-armed attack ships. As the Allied air offensive gathered weight, these ships mounted more and more defensive armament.

On 6 April, an attack was made on the steel mill at Caen, but cloud obscured the target. Typhoons had bombed Caen-Carpiquet airfield an hour before, but in that intervening time, the clouds thickened to 9/10 at 8,000 feet over Ouistreham. More sorties followed on 9 April and during the early afternoon of 13 April, but there was no contact with enemy.

A long flight of 300 miles (480km) was made to divebomb Brest airfield at 17.30 on 13 April. The squadron diary described it as 'this bright blue April evening.' The squadron departed on the attack from Predannack, escorted by the Portreath Wing, flew at zero feet for 12 minutes, then began climbing to cross the coast at Pontusval at 14,000 feet. The target was easily identified because of poor camouflage and a newly-made perimeter track. The aircraft went to echelon port, into their dives, and pulled out at 6,000 feet. Many hits were scored, some seen on the dispersal areas, the watch office and a nearby hangar. Flak followed the Whirlwinds back to the coast. Flying Officer Lee-White took off five minutes late, catching up over Guipavas, and Flying Officer Lovell became detached after bombing, so returned alone. For Max 'it was great fun, and there were FW190s taking off in all directions. By the time they got up to our height, we were halfway home again.'

Next day, six other pilots flew reconnaissance looking for E-boats which had attacked a British convoy off Falmouth.

The Norwegian destroyer *Eskdale* and British merchant ship *Stanlake* had been sunk. No E-boats were seen but a floating Lancaster was found, its crew in a dinghy, off the English coast. The Whirlwinds circled until the air-sea rescue organisation took over. At 15.30, five Whirlwinds led by Flying Officer Harvey attacked four ships, inflicting damage with bombs and cannon. Sergeant John Macaulay was last seen attacking the fourth ship, and the only news from him was a radio message heard by the leader of the escorting Spitfires, indicating he was going down to land. It was hoped that he got down safely on the Brest Peninsular. Two Junkers W34s were attacked, one being shot down by a Spitfire and one shot at by Harvey, with no results noted.

During the moon period of April, the squadron went hunting over France. Losses accumulated rapidly. Eddie Brearley was lost on 16 April, with no word whatsoever as to what happened. That day, Max led Yellow Section in the formation led by Jimmy Coyne, escorted by 12 Spitfires, looking for ships by day but, as Coyne noted, 'just gave the coastal guns some target practice.' On a night operation on the 17th, Rex King, Tim Harvey and Bas Abrams were lost. In Max's opinion, King was the best pilot in the squadron, and had received the DFM for shooting down a German without a gun-sight — no mean feat. In Jimmy Coyne's words, the flight was a miniature Commonwealth, with Max from Australia, Harvey from Ireland, King from the West Indies, Abrams from South Africa and Coyne from Canada.

Light flak and searchlights caught them, and Coyne, in a searchlight with volumes of light flak streaming along it at him, only escaped by 'ducking down behind a church steeple and a line of large trees.' He abandoned his search for trains, bombed the rail lines near Neuilly and headed home. Max was off to a side, and recognised King's aircraft caught in a cone of searchlights, with streams of flak coming up at him from several directions, and 'he didn't look as if he had a chance of getting through it.'

Like other pilots, Max thought flak more frightening at night, as each tracer shell can be seen clearly, streaming up, 'and it all seems to be coming so darn close to you. The only way to dodge it is to twist and turn and dive and climb and do anything else you can think of, but a Whirlwind is not easy to throw about with a load of bombs on.' He particularly disliked 'Flak Alley', the water gap between Alderney and the French mainland, as much flak would come from both sides when the aircraft passed through. However, he did find something to like about the night operations — each pilot was on his own, with no one else to worry about.

After the loss of King, Harvey and Abrams the squadron confined itself to training replacements and looking for ships around the Channel Islands. On the night of 18 April, Max flew with Flight Lieutenant Blackshaw on an unsuccessful search of the Channel Islands area. Four other pilots also found nothing. On the 20th, at night, Jimmy Coyne found five ships steaming north-east from Guernsey; he attacked and believed he scored a hit on one. Lee-White attacked two armed trawlers two miles (3km) south-east of St Marcouf, scoring a hit, despite moderate flak. Both pilots returned after midnight, and Geoff Warnes took Blackshaw, Simpson and Cotton out after Coyne's convoy, but could not locate it. Possibly the ships had gone into port at Alderney.

On 27 April, two RAF Mustangs on reconnaissance found a convoy off the Channel Islands, and 263 Squadron attacked with the Ibsley Wing of Spitfires as escort. 616 Squadron sent ten Spitfire VIs, and 504 Squadron provided twelve Spitfire V-Bs as anti-flak force.

Six Whirlwinds, led by Squadron Leader Warnes, with Flight Lieutenant Blackshaw leading the second flight of three, found the convoy of nine ships just south of Jersey, some 10 miles (16km) west of Corbiere Point. 504 Squadron attacked to keep down the heads of the flak crews, and each Whirlwind section picked a ship, then went for it. The bombs were fitted with three-second fuses, so the No. 2 in each section had to keep up with his leader. The six Whirlwinds kept together and attacked all at once. Flak flashed and splashed in the water all around the six planes, but only Flying Officer Lee-White was hit, in the tailplane, and even that was thought to be a hit from debris.

Warnes, Blackshaw and Max all bombed the same ship, a 1500 ton motor vessel, Lee-White hit an armed trawler, and Sergeant Simpson bombed a 'converted yacht' then hit an E-boat with cannon fire as the formation hurtled past out of gun range. Max was 'delighted to see great clouds of smoke and flame rising' from the ship he had hit. Three ships were on fire and were claimed sunk. German records showed the auxiliary minesweeping tender *Etienne Rimbert* M.4611 (197 tons) and the *Helma* (187 tons) were sunk.

On the way home, Warnes led them a little too close to Alderney, and the gunners there opened fire, dropping large shells into the water around the formation. They were used to being shot at on the way home, but not so accurately. One of the pilots was on his first operation, and after the business of attacking ships at sea-level, the flak was a bit too much. He opened up the engines and shot past the rest of the formation, leaving them well behind. Warnes had a difficult task in persuading him to return to his proper position.

On 28 April, Squadron Leader Warnes led Blackshaw, Simpson and Cotton again, escorted by the Exeter Wing, with a Czech squadron (310) close to the Whirlwinds. Warnes' aircraft would not start, so he took over that of Warrant Officer Tebbit, then Lee-White's had trouble, so only four set off on the strike against ships reported north of Sept Iles. The Spitfires went in alongside the Whirlwinds and strafed the ships to keep down the heads of the gunners, and the Whirlwinds attacked.

About half the ships were armed minesweepers or flak ships, and Flying Officer Pavlus of 310 Squadron was shot down by the intense light and heavy flak, going straight into the sea. A couple of others pulled away with damage after the firing pass over the convoy. Warnes and Blackshaw scored four hits on the rear-most ship, an M-class minesweeper, probably sinking it. Simpson hit an armed trawler with bomb and cannon, damaging it, and Max also damaged a mine-

An M-Class minesweeper under attack. (RAAF official)

sweeper with cannon and bombs. From Max's viewpoint, it seemed every calibre of gun aboard the ships was firing, and he saw a multi-barrelled light cannon blazing at him as he went for the minesweeper. 'The water between me and the plane in front was a mass of bullet splashes, and how we got through I don't know. I came in low, about 10 feet over the water, gave the ship a good long squirt with my cannon, released the bombs just before reaching it, and pulled up just in time to clear the mast. As I was going over the top, I felt a hell of a bump and looked out on the starboard wing to see a hole that must have been nearly two feet square. There was a large piece of metal sticking up on the wing, trying to pull it down, and I had to use both hands and my knee against the stick to hold the wing up.'

He was faced with a flight back over water to England, and then to Exeter. He could not keep up with the other three, so came along behind, with a Spitfire on either side. He later wrote that the Czech squadron leader saw his ship sink, as well as two more, so counted it a successful day. By the time he got to Exeter and landed, the strain of holding the starboard wing up was telling, and Max was quite weary.

Other shipping strikes were flown on 29 and 30 April, with no result. Another sweep was flown on 2 May to the Channel Islands area, but no ships were found. The squadron diary commented 'It seems the Casquets had again been reported as a tanker.' Next day, another operation with a minesweeper as target was flown, again with no ship found. However, six FW190s engaged, but the Spitfire escort protected the Whirlwinds, which were not attacked, though one Czech pilot of 310 Squadron was killed and another wounded, while Wing Commander Dolezal damaged a Focke-Wulf.

On 5 May, yet another fruitless search was made, this time for a minelayer, but only fishing vessels were found. The squadron landed at Bolt Head, in case anyone needed fuel, before going on back to Warmwell. Next day another unsuccessful search was made for a tanker with five small escort vessels, and the squadron noted that 'the Casquets were indeed found in this position.' Ever ready to shoot, the Alderney flak opened up with heavy accurate fire, but hit no one. Four Whirlwinds, with 129 and 504 Squadrons providing 16 Spitfires as escort, searched unsuccessfully on 11 May south of Jersey. On 14 May, seven aircraft, with the usual escort from the Ibsley Wing, made a very good approach to attack two destroyers and a motor vessel in Cherbourg Harbour. The squadron achieved surprise by climbing from sea-level to 12,000 feet into a cloud, from which they emerged out of the sunset and, partially obscured by the sun-reflected glow on the cloud, went into their dives, but were only able to claim a good bombing pattern with no hits. The flak had been slow to react.

Night sorties were flown on 15 May, after ships located by Flying Officer Lee-White and Jimmy Coyne, and Lee-White scored a hit. Squadron Leader Warnes took four Whirlwinds to attack, but it was not clear if hits were scored, and Flight Lieutenant Blackshaw was killed when he baled out too low, after his lights and radio failed. Later in the day, another strike against a motor vessel and two destroyers west of the Casquets was attacked head-on by two FW190s. But when the 190s came back for a second pass, Lee-White and Coyne turned into them and, bombs still on their wings, hit both with good shooting. The Germans escaped, despite the escorting Spitfires. Other uneventful reconnaissances were flown during the next few days.

Max had been on a week's leave, which he spent with the Warrens in Maidstone. 'It was spring, and Kent looked really green and beautiful. It is known as "The Garden of England" and I think it is one of the prettiest parts, but I still like Bonnie Scotland best.' He played golf with two of the young

Doctors Warren, went swimming, played tennis, went to dances and generally enjoyed a relaxing time with friends. Max was next on operations on the night of 18/19 May, with no enemy located.

Eddie Brearley had been missing since April, and now his body was washed ashore, so his funeral followed that of Blackshaw. Max wrote that funerals were frequent, but nobody got morbid about them, instead going into Weymouth to brighten things up a bit.

On the night of 21 May, the squadron was having what he described as 'a fairly merry time in the Mess', when Operations rang to order a strike on ships located off Cherbourg. Squadron Leader Warnes looked around and took those officers present with him.

What had happened was that after the recent period of intense flying, Squadron Leader Warnes had extracted a promise from 10 Group that the squadron would not be called on during the night. The squadron had an old friend near Charmy Down, a butcher. The Miles Magister was sent there and returned with steaks, chops, kidneys and other good food, which the Mess Sergeant prepared. Jimmy Coyne said that they had 'just completed this wonderful dinner, and were about to begin the serious party, when Group called.' Black coffee was given to the pilots of the five flyable aircraft, and they went.

Taking off at 25 minutes past midnight, Warnes led Joe Holmes, Max Cotton, Jimmy Coyne and Arthur Lee-White out after the convoy believed to be steaming from Cap de la Hague to Cherbourg. Warnes left first, with the others to follow as soon as possible. He found the ships, then circled and radioed directions to the others, before going in to bomb a 3,500 ton motor vessel escorted by four armed trawlers. He pulled away after bombing, circled and watched the next attacks.

After crossing the English coast, and staying at sea-level for 15 minutes, Max climbed to 3,000 feet. It was a clear moonlit night, and he could see Alderney plainly ahead of him. He heard Warnes call that he had 'found the bastards, three miles east of Cap de la Hague.' Then Max saw the ships. Holmes attacked next, hit the same one attacked by Warnes, and started a fire on board the 1,500 tonner. The ships were in sight of the entrance to the mole at Cherbourg, and Jimmy Coyne thought the crews were preparing to dock and looking forward to relaxation in the local estaminets. The guns had probably been secured, allowing the first attackers to go in with no retaliation.

Max also saw the four escorts, 'carrying more guns than you could count. I got them lined up directly into the moon and went in hell for leather. The CO had encountered no flak, Joe Holmes just a little, and when it came to my turn they were all ready and waiting. As soon as I dropped my bombs, shells started pouring up at me. It looked like so many red-hot coals carving their way through the sky. It was coming bloody close, and no matter how much I twisted and turned, I couldn't get away from it. Suddenly there was a terrific bump and an explosion, and my kite rocked like blazes.'

A shell had hit the rear of the fuselage, cutting the airspeed indicator tube and the elevator trim cable. A machinegun bullet had clipped through the cockpit canopy about two feet (60cm) behind Max's head.

Holmes and Max attacked the same ship as Warnes, and their combined efforts finished it off. Jimmy Coyne sank an armed trawler, and Lee-White went for one of the remaining three. By this time, the gunners were well and truly stirred up, and hit him some 400 yards (metres) away, so that his starboard engine was on fire, but he went on to bomb. Pulling off the target, Lee-White radioed that he was about to bale out over Queiqueville, but the slipstream seemed to be dealing with the fire, so he changed his mind and flew back to Warmwell, where he landed with the engine burning well, but the fire-tender was on hand to assist.

With no airspeed indicator, Max's immediate problem was a safe landing at night. His first approach was made 'like a bat out of hell', so he went round again. After several attempts, he got down safely. 'All the others laughed like hell when they saw Lee-White's and my machine, all full of holes, thought it a grand joke. But the trip was well worth it. After that we went back to the bar (it was 2 a.m.) and did another hour's steady drinking, as though the drinking had not been interrupted.' The RAF of 1939-45 carried on the traditions of the RFC 1914-18, in which great danger and narrow escapes were passed off with understatement and humour, mixed with youthful enthusiasm. Those who did not understand sometimes thought this attitude unthinking and callous, but it was the best way for a wartime organisation to carry on through the daily hazards.

The squadron diary described Max's aircraft as Category B damage, 'with a fuselage like a pepper pot. He made a good landing after four attempts without ASI and with rudder partly jammed. Neither aircraft had hydraulics; wheels and flaps answered to the emergency air bottle. This operation was the most successful, exciting and spectacular of our night operations to-date.' Max merely noted in his log book, '2 escort ships sunk, one merchant ship damaged, got shot-up again.' Post-war research showed the Dutch ship *Oost Vlandered* of 412 tons was sunk, but there were no German records of escort ship losses.

Another convoy battle ensued on 23 May. A reconnaissance Spitfire which found it had been shot down. The ships were in line astern, east of St Peter Port, ready for attacks from north to south. The Whirlwinds, with the Ibsley Wing, flew between Guernsey and Heron, and heavy flak, to attack. Flight Lieutenant Holmes was leading with Max as his No.2, Jimmy Coyne and Ridley. Holmes put his own bombs into the waterline of a Dutch coaster while the following three damaged an armed trawler.

Jimmy Coyne recalled, 'on the run-up to the convoy we were heavily engaged by all types of guns from both Sark and Guernsey as well as the ships and harbour defences. The surface of the water looked for all the world like it was being pounded by monstrous hailstones. So heavy was the firing, I thought there was no way we could come through this one unscathed. Joe and Max were slightly ahead and to the right. I saw Joe's bombs make a direct hit on the largest ship in the convoy. Max was in his usual solidly correct position but his bombs seemed to have been dropped slightly prematurely and I saw them explode very near to the stern. Ken and I attacked the next largest ship, and had direct hits reported by the Spits.'

Max himself described the attack. 'We went tearing in at full throttle and got a very warm reception. The water all around us was fairly boiling with gunfire. Both the ships and shoreline were throwing up everything they had. I had just released my bombs when I felt the not unusual thump of a cannon shell. This time they got me through the starboard petrol tank, and 25 gallons poured out in a few seconds. It was a self-sealing tank and "mended itself" when there were still 15 gallons left. Even so, that did not leave me much to get back to Warmwell.' He climbed to 5,000 feet, to have enough height to return on one engine. The starboard tank read '0' for the last five minutes of the flight, and when he taxied in there was only enough left, as he said, 'to fill a cigarette lighter.' Max was hit in the starboard wing with a 20mm shell, and the squadron diary recorded his return to Warmwell 'with his third aircraft Category B damage for Westlands'.

Max had been hit three times in the previous four shipping attacks, and was dubbed 'Flak Attractor'. He wrote that seeing the water 'boiling all around me shook me a bit for the start, but when I had been through a few times, I got used to it. However, it *was* rather unhealthy.' This was his last journal entry.

Jimmy Coyne, and the squadron, 'wondered how much longer his luck could hold out.'

The supply of Whirlwinds was becoming exhausted, and on 25 May the squadron was given the bad news that they would be re-equipped with another type of aircraft: Hurricane IVs. By now the Hurricane was well past its prime, had not been allowed on fighter sweeps since late 1941, but was still employed on fighter-bomber attacks. Flak and fighters inflicted severe losses on the unfortunate pilots. But 263 was reprieved, as 137 Squadron was to hand over its Whirlwind aircraft and be equipped with a different type. However, it was obvious the Whirlwind was to be replaced, and later 263 would fly Typhoons.

On 29 May, six Whirlwinds, with Max leading the second flight of three, went off with the Ibsley Wing to attack a convoy of 15 ships off Barfleur, which proved to be the French fishing fleet. On other days of the month, more practice flights were made, training the new arrivals. Increasing emphasis by Fighter Command on anti-shipping activity is shown by the following:

March: 179 sorties, making 28 attacks, for one aircraft lost; April: 723 sorties, 97 attacks, 16 aircraft lost; May: 953 sorties, 78 attacks, nine aircraft lost; June: 1078 sorties, 127 attacks, 13 aircraft lost.

The night attacks resulted in little shipping movement along the coast in the moon periods.

The June weather was mostly poor, with 19 days of frontal

activity—rain, wind and low clouds. There were no operations flown in the first five days of the month, and nothing was achieved in operations to 12 June. During this time, Max flew several weather tests, camera-gun combats, formation practices and other training sorties. There were other matters of note, as on 2 June DFCs were awarded to Flight Lieutenants Holmes and Coyne, followed on 13 June by the award of more DFCs to Flying Officer A. Lee-White and Pilot Officer M.T. Cotton, plus a DSO to Squadron Leader Geoff Warnes. Flight Sergeants R.A. Hollamby and H.O. Watts were Mentioned in Despatches for their work on the ground in the squadron. A squadron party was held at the 'Golden Lion' in Weymouth, a 'magnificent party, magnificently enjoyed', said the squadron diary.

The citation for Max's DFC read:

> Pilot Officer Cotton has taken part in numerous and varied operational missions, some of which have been completed at night. His sorties have included bombing attacks on enemy airfields, railway objectives, and against shipping in convoy. On three occasions, his aircraft has been severely damaged by anti-aircraft fire, but he did not fail to fly back to this country to land safely. Pilot Officer Cotton has displayed complete disregard of intense enemy opposition and has invariably pressed home his attacks with great determination.

There was no flying on the 14th, but a reconnaissance was flown on 15 June, and found enemy ships. Flying Officer Lee-White led Sergeant G.A. Wood, Max and Flight Sergeant K. Ridley, with 16 Spitfires from 504 and 616 Squadrons as anti-flak and escort. Squadron Leader 'Laddie' Lucas, CO 616 Squadron, led the formation at sea-level for the north of Guernsey. His Spitfires were to attack as anti-flak, while 504 Squadron escorted and covered the Whirlwinds.

'Those shipping strikes were most unpleasant affairs,' recalled Lucas. 'The enemy vessels plying in that area were invariably heavily armed and the flak from their AA guns was most lethal. Moreover, they got plenty of practice. We rendezvoused with the Whirlwinds about five miles off the Dorset coast and set course southwards at nought feet well under enemy radar cover. It was a grey, overcast day, but with excellent visibility.'

When some three miles (5km) north-east of Sark, Lucas turned left, and immediately saw two sections of two ships ahead, steaming to Cap de la Hague. 'The ships were sighted when we were probably six or eight miles away; they probably saw us down on the water four or five miles off. With these conditions there was no way that the attack could be a surprise—it was a question of heading straight at the vessels with the Whirlwinds spreading out as they made their run-in, flat out, at the target.'

The ships were four minesweepers, and Laddie climbed to 400 feet and dived from 1,000 yards (metres) to attack. The ships immediately opened fire, as did flak on the nearby islands, and Flying Officer Bob Sim RNZAF was hit, his Spitfire burst into flames and he passed over the convoy before going into the water on the far side. Sim had survived

Pilot's eye view of a low level attack on a German convoy, similar to those made by 263 Squadron. (RAAF official)

the hard combats of the siege of Malta in 1942. The Spitfires sprayed all four ships with cannon and machinegun fire and then were past, with the Whirlwinds coming in close behind.

Flying Officer Lee-White and Sergeant Wood attacked the third ship below mast-height, and smoke was seen from it after the bombs exploded.

Max's bombs were seen to splash into the waterline of the leading ship but, as he hurdled the minesweeper, his aircraft was hit in the cockpit area by a 40mm shell, and went at full speed into the sea. There was no chance of survival. Laddie Lucas saw the crash: 'As the aircraft hit the sea there was a hell of a lot of spray and smoke, but we couldn't wait around to establish its fate. The Whirlwinds turned left-handed towards the Cherbourg peninsula and then north for home. We withdrew uneventfully from the attack. No one cared much for these anti-shipping operations, as the cross-fire from the ships' light AA was always intense.' German records confirm M-class minesweeper M483, 750 tons, was sunk.

Ridley's Whirlwind had Category B damage to the fin and rudder, and the others returned unscathed.

The squadron diary stated that:
> P/O Max Cotton came to the squadron in July 1942. From the first he proved himself to have an extraordinary flair for accurate bombing. Both in practice and in action he rarely missed his direct hit. As a pilot and a leader his matter-of-fact calmness before, during and after operations made him invaluable to the squadron. As a person, his courtesy and frank and open nature and constant good humour endeared him to all who knew him.

Squadron Leader Geoff Warnes wrote to Max's father:
> You will have heard from the Air Ministry that Max is missing, believed killed. This morning, in attacking a German ship, he was hit by flak and went straight into the sea. He would never

know what happened, but before being hit he had dropped his bombs and as usual hit the target.

There is not much I can say as I do not want to appear verbose or insincere, my full thoughts are neither.

Max was a great chap who was always terrifically keen and never was a DFC better earned than his.

The squadron and the United Nations can ill afford the loss of such a fine fellow. Always very popular, he is greatly missed, but, of course, what is our loss compared with yours?

The whole squadron, officers, NCOs and even groundcrew, have asked me to send to you their sympathy in this very sad time.

Postscript

Jacques Andrieux, who shot down one of the FW190s attacking Max on 28 February 1943, survived the war, ending it as CO of 341 (Alsace) Squadron with a DFC and Bar. He remained in the French Air Force, retired as a General and, at time of writing, lived on a Paris street named for Nungesser and Coli, lost trying to cross the Atlantic.

Squadron Leader Warnes DSO DFC, the flight commanders and other pilots were posted away from 263 Squadron soon after Max Cotton's death. A pilot posted into the squadron, who remained to fly Typhoons with it, was 409257 Flying Officer Robert Bruce Tuff RAAF. Born at Seddon, Victoria on 11 July 1922 Tuff was educated at Bendigo High School, worked at the State Electrical Commission and Taxation Office as a junior clerk, joined the RAAF on 21 June 1941, embarked for overseas on 16 June 1942—just short of his 21st birthday—and eventually was posted to 263 Squadron. In early 1944, Geoffrey Warnes returned as CO; the squadron had been equipped with Hawker Typhoons. On 2 February 1944, while returning from France, Warnes' engine failed and he went down into the cold water. The squadron circled, and Warnes could be seen in difficulty, trying to get into his dinghy. The weather was bad, and all knew Warnes had little chance if he could not get out of the water. Bob Tuff said he was going to bail out to help. He ignored calls to dissuade him, and was seen to parachute into the water not far from Warnes. The weather closed in; the Typhoons had to leave for England; neither man was seen again. The decision to go to Warnes' assistance was not made in heat of battle, adrenalin surging and reactions on the instant, but was a result of an assessment of the chances for survival of a fellow pilot in desperate circumstances. This selfless action by Bob Tuff has been included in at least two other books, but is worthy of inclusion in this brief account of part of the life of 263 Squadron, and of two young Australians who served in the unit.

Appendix

Aircraft flown on operations by Pilot Officer MT Cotton DFC RAAF:

Westland Whirlwind	P7108	18	sorties
	7013	5	sorties
	6971	3	sorties
	7000	3	sorties
	7062	3	sorties
	6974	1	sortie
	6981	1	sortie
	7052	1	sortie
	7057	1	sortie
	7089	1	sortie
	7117	1	sortie

Pilot Officer Max Cotton flew 38 operational sorties, and had a total of 465 hours 50 minutes flying time (log book).

Max Cotton DFC

Aircraft flown on notable operations:

Date	Serial number	Operation Detail
14 December 1942	P7052	Combat with two FW190s
26 February 1943	P6971	Dive-bomb Maupertus
27 February 1943	P6971	Dive-bomb Maupertus
28 February 1943	P6971	Maupertus—FW190 on tail
14 March 1943	P7108	Dive-bomb Maupertus
13 April 1943	P7013	Dive-bomb Brest airfield
17 April 1943	P7103	Night operation; 3 shot down
27 April 1943	P7108	Shipping strike
28 April 1943	P6981	Shipping strike—hole in wing
21 May 1943	P7108	Shipping strike—hole in tail
23 May 1943	P7089	Shipping strike—hole in tank
15 June 1943	P7000	Shipping strike—killed

Footnotes

That brand of courage
1. Combat report: PRO Kew II N 40/d7/2
2. AWM 220 Item 6, Campaign in France and the Low Countries
3. ibid.
4. ZG2 crew identification via John Vasco
5. AWM 220 Item 6
6. ibid.

Bader's Bus Company
1. AWM 220 413/4/103 Translation VII/148
2. ibid. Translation VII/144
3. ibid. Translation VII/89
4. ibid. Translation VII/142
5. ibid. Translation VII/143
6. ibid. Translation VII/162
7. ibid. Translation VII/130
8. ibid. Translation VII/164
9. ibid.

Take that off my tax!
1. RAAF Historical Office files; RAAF Personnel records
2. Letter, Tom Warren/author 7 Nov. 88
3. ibid.
4. ibid.

We never met his equal
1. 418 Squadron history
2. 418 Squadron ORB
3. Letter, JRF Johnson, 18.11.88
4. ibid.
5. AWM 220, RAF 1939-45 Vol. V, 'ADGB'
6. Letter, Tom Anderson, Feb. 89
7. 418 Squadron history
8. Letter, Howell, Nov. 88

Bibliography

Blitzed! by Victor Bingham, Air Research Publications, 1990.
Der Adler, various issues, 1941-44.
Die Ritterkrueztrager der Luftwaffe, Band I & II, Ernst Obermaier, Verlag Dieter Hoffman, Mainz, 1976.
Duel for the Sky by Christopher Shores, Blandford Press, UK, 1985.
Duel under the Stars by Wilhelm Johnen, William Kimber, 1957.
Fighter! by Werner Held, Arms & Armour Press, UK, 1979.
Fighter Pilot by Paul Richey, Hutchinson, 1941.
Fighter Aces by Christopher Shores, Hamlyn, 1975.
JG26, Geschichte Eines Jagdgeschwaders by Josef Priller, Motorbuch Verlag, Stuttgart, 1980.
Kampfgeschwader 'Edelweiss' by Wolfgang Dierich, Ian Allen, 1975.
Pictorial History of the Luftwaffe by Alfred Price, Ian Allen, 1969.
RAAF over Europe, edited by Frank Johson, Eyre & Spottiswoode, UK, 1946.
Reach for the Sky by Paul Brickhill, Collins, 1954.
2nd Tactical Air Force by Christopher F. Shores, Osprey, UK, 1970.
Six Months to Oblivion by Werner Gerbig, Ian Allen, 1975.
Swastika in the Air by Karl Bartz, William Kimber, 1956.
The Big Show by Pierre Clostermann, Chatto & Windus, 1951.
The Defence of the Reich by Werner Held & Holger Nauroth, Arms & Armour Press, UK, 1982.
The First and the Last by Adolf Galland, Methuen & Co., 1955.
The German Jets in Combat by Jeffrey Ethell & Alfred Price, Jane's, 1979.
The Luftwaffe War Diaries by Cajus Bekker, Macdonald & Co., 1967.
The Messerschmitt Me262 Combat Diary by John Foreman & S.E. Harvey, Air Research Publications, 1990.
The Mighty Eighth by Roger A. Freeman, Macdonald, London, 1970.

The Rise and Fall of the Luftwaffe by David Irving, Wiedenfeld & Nicolson Ltd, 1976.
Tumult in the Clouds by James A. Goodson, William Kimber, 1983.
Whispering Death by Neville Parnell, Reed Books, Sydney, 1980.
Wing Leader by J.E. Johnson, Chatto & Windus, 1956.